The Creation/Evolution Controversy

THE MAGILL BIBLIOGRAPHIES

The American Presidents, by Norman S. Cohen, 1989
Black American Women Novelists, by Craig Werner, 1989
Classical Greek and Roman Drama, by Robert J. Forman, 1989
Contemporary Latin American Fiction, by Keith H. Brower, 1989
Masters of Mystery and Detective Fiction, by J. Randolph Cox, 1989
Nineteenth Century American Poetry, by Philip K. Jason, 1989
Restoration Drama, by Thomas J. Taylor, 1989
Twentieth Century European Short Story, by Charles E. May, 1989
The Victorian Novel, by Laurence W. Mazzeno, 1989
Women's Issues, by Laura Stempel Mumford, 1989
America in Space, by Russell R. Tobias, 1991
The American Constitution, by Robert J. Janosik, 1991
The Classical Epic, by Thomas J. Sienkewicz, 1991
English Romantic Poetry, by Bryan Aubrey, 1991
Ethics, by John K. Roth, 1991
The Immigrant Experience, by Paul D. Mageli, 1991
The Modern American Novel, by Steven G. Kellman, 1991
Native Americans, by Frederick E. Hoxie and Harvey Markowitz, 1991
American Drama: 1918-1960, by R. Baird Shuman, 1992
American Ethnic Literatures, by David R. Peck, 1992
American Theatre History, by Thomas J. Taylor, 1992
The Atomic Bomb, by Hans G. Graetzer and Larry M. Browning, 1992
Biography, by Carl Rollyson, 1992
The History of Science, by Gordon L. Miller, 1992
The Origin and Evolution of Life on Earth, by David W. Hollar, 1992
Pan-Africanism, by Michael W. Williams, 1992
Resources for Writers, by R. Baird Shuman, 1992
Shakespeare, by Joseph Rosenblum, 1992
The Vietnam War in Literature, by Philip K. Jason, 1992
Contemporary Southern Women Fiction Writers, by Rosemary M. Canfield Reisman and Christopher J. Canfield, 1994
Cycles in Humans and Nature, by John T. Burns, 1994
Environmental Studies, by Diane M. Fortner, 1994
Poverty in America, by Steven Pressman, 1994
The Short Story in English: Britain and North America, by Dean Baldwin and Gregory L. Morris, 1994

Victorian Poetry, by Laurence W. Mazzeno, 1995
Human Rights in Theory and Practice, by Gregory J. Walters, 1995
Energy, by Joseph R. Rudolph, Jr., 1995
A Bibliographic History of the Book, by Joseph Rosenblum, 1995
The Search for Economics as a Science, by the Editors of Salem Press (Lynn Turgeon, Consulting Editor), 1995
Psychology, by the Editors of Salem Press (Susan E. Beers, Consulting Editor), 1995
World Mythology, by Thomas J. Sienkewicz, 1996
Art, Truth, and High Politics: A Bibliographic Study of the Official Lives of Queen Victoria's Ministers in Cabinet, 1843-1969, by John Powell, 1996
Popular Physics and Astronomy, by Roger Smith, 1996
Paradise Lost, by P. J. Klemp, 1996
Social Movement Theory and Research, by Roberta Garner and John Tenuto, 1996
Propaganda in Twentieth Century War and Politics, by Robert Cole, 1996
The Kings of Medieval England, c. 560-1485, by Larry W. Usilton, 1996
The British Novel 1680-1832, by Laurence W. Mazzeno, 1997
The Impact of Napoleon, 1800-1815, by Leigh Ann Whaley, 1997
Cosmic Influences on Humans, Animals, and Plants, by John T. Burns, 1997
One Hundred Years of American Women Writing, 1848-1948, by Jane Missner Barstow, 1997
Vietnam Studies, by Carl Singleton, 1997
British Women Writers, 1700-1850, by Barbara J. Horwitz, 1997
The United States and Latin America, by John A. Britton, 1997
Reinterpreting Russia, by Steve D. Boilard, 1997
Theories of Myth, by Thomas J. Sienkewicz, 1997
Women and Health, by Frances R. Belmonte, 1997
Contemporary Southern Men Fiction Writers, by Rosemary M. Canfield Reisman and Suzanne Booker-Canfield, 1998
Black/White Relations in American History, by Leslie V. Tischauser, 1998
The Creation/Evolution Controversy, by James L. Hayward, 1998
The Beat Generation, by William Lawlor, 1998
Biographies of Scientists, by Roger Smith, 1998
Introducing Canada, by Brian Gobbett and Robert Irwin, 1998
Four British Women Novelists: Anita Brookner, Margaret Drabble, Iris Murdoch, Barbara Pym, by George Soule, 1998

The Creation/Evolution Controversy

An Annotated Bibliography

James L. Hayward

Magill Bibliographies

The Scarecrow Press, Inc.
Lanham, Md., & London
and
Salem Press
Pasadena, Calif., & Englewood Cliffs, N.J.
1998

SCARECROW PRESS, INC.

Published in the United States of America
by Scarecrow Press, Inc.
4720 Boston Way
Lanham, Maryland 20706

4 Pleydell Gardens
Kent CT20 2DN, England

British Library Cataloguing in Publication Information Available

Library of Congress Cataloging-in-Publication Data

Hayward, James L., 1948–
The creation/evolution controversy : an annotated bibliography / James L. Hayward.
p. cm. — (Magill Bibliographies)
Includes indexes.
ISBN 0-8108-3386-7
1. Evolution (Biology)—Bibliography. 2. Evolution—Bibliography. 3. Creationism—Bibliography. I. Title. II. Series.
Z5322.E9H39 1998
[QH366.2]
016.5768—dc21
98-3138
CIP

ISBN 0-8108-3386-7 (cloth : alk. paper)

™ The paper used in this publication meets the minimum requirements of American National Standard for Information Sciences—Permanence of Paper for Printed Library Materials, ANSI Z39.48–1984.
Manufactured in the United States of America.

To my father and mother,
James L. Hayward, Sr. and Jane B. Hayward,
who taught me to value
honesty, fairness, and truth.

CONTENTS

Preface

This collection of 447 annotated references will provide students, teachers, lawyers, writers, historians, scientists, clerics, and other interested persons with an overview of the literature dealing with the creation/evolution controversy. The 54 references in Chapter 2 highlight influential volumes published between 1543 and 1980—classic works that informed the views of later writers. The remaining 393 entries in Chapters 3 to 7 feature books published from 1981 to 1996.

The year 1981 seemed like an appropriate one at which to begin the regular listings. During that year (1) the Creation-Science Research Center fought a court battle with the California State Board of Education to keep a 13-year-old boy from being taught evolution in a public school science class, (2) a law mandating equal time for creation-science in the public schools was passed in Arkansas, and (3) a similar equal treatment law was passed later that year in Louisiana. Americans were once again fighting over issues made famous by the Scopes "monkey trial" waged some 56 years earlier.

This bibliography is not exhaustive, but it is representative. Many of the listed works address the creation/evolution debate directly, despite that fact that in some of the sources this topic takes a back seat to other concerns. Children's books have not been included. Readers interested in a comprehensive annotated listing of earlier creationist books should consult Tom McIver's *Anti-Evolution: A Reader's Guide to the Writings Before and After Darwin* (1988) listed in Chapter 2; similarly, David Hollar's *The Origin and Evolution of Life on Earth* (1992), listed in Chapter 7, consists of an extensive set of annotated references to sources on evolutionary biology.

Some of the listed works I have read cover to cover, others I have carefully perused, and still others I have summarized with the help of reviews published in *Creation Research Society Quarterly, Origins,* and *Perspectives on Science and the Christian Faith* (formerly *Journal of the American Scientific Affiliation*). My thanks to the writers of these reviews and my apologies to the authors of books I have misrepresented or overlooked—these were sins of omission, not of intent.

In Chapters 3 to 7 books are classified as either theistic (assuming the activity of a deity) or nontheistic (ignoring or denying the possibility of a deity). These terms were sometimes difficult to apply, especially in cases of ambiguity or multiple authorship. Undoubtedly I made some misjudgments.

Except for the historical references in Chapter 2, I have used the present tense to refer to each author's occupation and status, even though these may have changed since publication of his or her book. If more than one book by an author is listed within a chapter, information about the author generally appears in only the first annotation.

Many people have contributed to this project, either directly or indirectly. My students, both graduate and undergraduate, serve as constant motivators—they ask good questions for which they expect honest answers. Gary Land critiqued an early draft of Chapter 1, Warren Johns offered helpful suggestions on Chapter 2, and Tom Goodwin provided thoughtful insights into a variety of books. My department chair, John Stout, furnishes a work environment that favors maximum productivity, personal growth, and creativity. A. J. Sobczak of Salem Press gave outstanding editorial advice and kept his cool each time I pushed back a completion date. Finally, my wife, Cheryl, and my daughter, Shanna, continued to be my best friends.

James L. Hayward
Berrien Springs, Michigan
September 1997

Chapter 1

INTRODUCTION TO THE CREATION/EVOLUTION CONTROVERSY

In 1925, the Tennessee legislature passed the Butler Act, which prohibited the teaching of evolution in the state's public schools. John T. Scopes (1901-1970), a high school science teacher in the town of Dayton, agreed to violate the new statute to give the American Civil Liberties Union a chance to test its constitutionality. The ensuing legal battle created an international stir, due partly to the famous attorneys who argued the case—William Jennings Bryan (1860-1925) for the prosecution and Clarence Darrow (1857-1938) for the defense—and to journalist H. L. Mencken's (1880-1956) acerbic descriptions of the trial. The defense lost and Scopes was fined, though ultimately the conviction was overturned on a technicality. The Butler Act remained on the books for more than four decades.

Years later the creation/evolution issue is once again making headlines. Preachers exhort. Parents protest. School boards act. Courts deliberate. Legislators battle. Writers warn. Scientists cry foul. The Scopes trial is played out countless times across a nation experiencing increasing political, scientific, and religious diversity.

Ironically, as one surveys the vast creation/evolution literature, it becomes apparent how impossible it is to categorize most people as simply "creationists" or "evolutionists." Many "creationists" believe in the occurrence of at least some evolutionary change, while some "evolutionists" posit the existence of a divine creator. Moreover, many evolutionary scientists are devout Jews, Christians, Hindus, and Muslims, while still others entertain the possibility of a creator but profess no formalized religious faith. Thus, it is somewhat misleading to speak of the "creation/evolution controversy" as if only two sides were represented. It is more accurate to view the conflict as one between "theists," who believe in some form of supernatural influence on the history of life and the universe, and "nontheists," who view this history from a purely naturalistic perspective. Consequently, the bibliographic entries in Chapters 4 to 7 are categorized according to these two perspectives.

This introductory chapter provides a brief history of evolutionism (both theistic and nontheistic) and creationism, examines some of the disagreements and challenges impacting each of these belief systems, and defines some basic terminology.

EVOLUTION AND EVOLUTIONARY THOUGHT

The creation/evolution controversy is centered around a philosophical line of reasoning known as the *argument from design.* William Paley (1743-1805), an English theologian of the late eighteenth and early nineteenth centuries, was the most influential proponent of this argument. He began his famous 1802 treatise on the topic, *Natural Theology: Or, Evidences of the Existence and Attributes of the Deity, Collected from the Appearances of Nature,* with an analogy: If a person was to find a watch in a field, he would assume that the watch had a designer and maker. A watch, after all, is a very complicated object. It is unlikely that such an object could arise by chance. Moreover, every part of the watch is necessary for its proper function. Remove one part and it will cease to function. Thus, the watch had to have been carefully planned and manufactured. As one looks at the world and its inhabitants, wrote Paley, one sees that, like the watch, they are very complicated. Therefore, the world and its inhabitants must have had a designer.

A young English naturalist, Charles Darwin (1809-1882), had earned a degree in theology from Cambridge University 29 years after publication of Paley's *Natural Theology.* Darwin liked reading Paley and was impressed by his arguments. Observations made during a five-year, round-the-world voyage (1831-1836), though, led him to believe that, although organisms certainly showed evidence of design, God was not the designer. How could a just and loving God have designed anything like the ichneumon wasps, he wrote, who slowly devour their prey alive? Darwin began to think about alternatives to Paley's argument (Desmond and Moore, 1991).

Darwin's ruminations led to publication of his 1859 classic, *On the Origin of Species by Means of Natural Selection,* one of the most influential books ever written. Darwin argued two main points in his book: first, that living things have changed significantly from their origin; and second, that this change has occurred as the result of a process he called *natural selection*. Other people had proposed theories of biological change before Darwin, but he was the first to publish a convincing mechanism by which such change could occur.

According to Darwin's theory, natural populations of organisms exhibit tremendous heritable variation. They also tend to produce more offspring than can survive. Offspring that do survive and reproduce are those variants best suited by heredity to live in the local environment. This differential reproductive success he called natural selection. Over time, as environmental conditions change, the types of organisms best suited to these conditions also gradually change. According to Darwin, this is how evolution works.

Actually, Darwin avoided use of the word "evolution" in his writings until close to the end of his life. During the nineteenth century this word implied progress, and Darwin was uncomfortable with this notion as applied to his theory. In fact, in one of his diary entries he tells himself to "Never say higher or lower" when referring to the outcome of biological change. It was social theorist Herbert Spencer (1820-1903) who first used the word "evolution" to refer to Darwin's theory. Spencer's designation became so popular that, ultimately, even Darwin himself decided he could no longer avoid its usage (Gould, 1977).

The first edition of *Origin of Species* sold out its first day, but the controversial book received a mixed review. Some people, like Spencer and biologist Thomas H. Huxley (1825-1895), became immediate supporters of Darwin's views. Others, like Harvard zoologist Louis Agassiz (1807-1873) and British physicist Lord Kelvin (1824-1907), found Darwin's thesis unconvincing. Many nonscientists opposed Darwinism because they thought it might upstage religion and morality. By the close of the nineteenth century, however, most of the scientific world had embraced some form of evolutionary theory, as had many mainline Christian denominations.

Despite its widespread acceptance, Darwin's particular brand of evolutionary theory has faced a number of challenges within the scientific community. One such challenge was the *theory of acquired characters*, popularized by French biologist Jean-Baptiste Lamarck (1744-1829). Lamarck defended the notion that evolution proceeded as a result of the acquisition of desirable traits during the lifetimes of individual organisms. These acquisitions were then passed along to their offspring. Thus, mused Lamarck, when food became scarce, early short-necked giraffes found it necessary to stretch higher and higher to browse for leaves. Because of this stretching, the offspring of these giraffes were born with longer necks than their parents. Evolutionists, including Darwin, effectively discounted the more extreme forms of Lamarckism, but occasionally, even today, Lamarckian-like ideas are seriously entertained. For example, some microbiologists think that bacteria may undergo Lamarckian-like evolutionary changes in response to antibiotics.

A second challenge arose during the early twentieth century, when many geneticists became convinced that natural selection, acting on genetic variation, played a negligible role in evolutionary change. They argued instead that evolution occurred only as a result of new mutations. An extreme variation on this theme was promoted during the 1940's by geneticist Richard Goldschmidt (1878-1958). According to Goldschmidt, the evolution of one species to another could not occur through a gradual accumulation of small mutations. Instead, major evolutionary changes required sudden, large-scale mutations resulting in the production of "hopeful monsters" which were occasionally selectively favored. Few biologists found Goldschmidt's theory appealing, however, for it lacked convincing supportive evidence.

A third, more recent argument centers around the question of molecular diversity. Many evolutionists have suggested that the differences in biological molecules exhibited by different organisms have resulted from natural selection. Defenders of the so-called *neutral theory of molecular evolution*, however, prefer to think that most molecular changes that occur in organisms are selectively neutral. Much evolution, they believe, is simply the result of these nonselective changes. Neutral evolutionists are not necessarily opposed to the concept of natural selection, only to the position that natural selection is the most significant force in molecular evolutionary change.

A fourth, contemporary debate concerns evolutionary tempo. According to Darwin, evolution occurs slowly and gradually. This view continues to be shared by many traditional evolutionists. A new breed of younger evolutionists, particularly paleontologists, however, believes that evolution proceeds in a more episodic than gradualistic fashion. Defenders of this perspective, called the *theory of punctuated equilibrium*, note that in many cases little change seems to occur in fossils found through significant portions of the geologic column. Then, suddenly, new body types appear. This evidence is interpreted to suggest that evolutionary changes occur rapidly (too rapidly for fossilization to be likely) followed by long periods of little or no change (during which time fossilization is likely). Although "punctuationalists" reject Darwinian gradualism, they continue to embrace natural selection as an important directive force in evolution.

Despite these squabbles over the particulars of the evolutionary process, Darwin's concept of evolution by natural selection remains the centerpiece of most theories of biological change. Since the time of Darwin, however, numerous discoveries have been made in the field of genetics that have filled in details about how change occurs, details that Darwin could not have known. This contemporary melding of genetics with classical Darwinian theory is referred to as *neo-Darwinism* or the *synthetic theory of evolution* (Price, 1996).

The term "evolution" carries numerous meanings today, all of which denote action and change. In its broadest sense, evolution refers to processes through which the universe, solar system, planet, or organisms have changed since their origins. Sometimes the term "evolution" is used to designate both events of origin *and* post-origin processes; but to avoid confusion it is best to reserve the word "evolution" to refer to post-origin processes. Unless otherwise noted, this book will use the term "evolution" to refer to post-origin processes of biological change.

Some important variations on the meaning of the word "evolution" are as follows:

1. Evolution is a change in gene frequencies from one generation to another. This process is also called *microevolution.*
2. Evolution is a change from one species to another. This process is also called *speciation.*
3. Evolution is the development of new adaptive features in organisms. This process is also called *macroevolution.*
4. Evolution is the development of major new groups of organisms. This process is also called *megaevolution.*
5. Evolution refers to the history of life from its origin to the present. This process is also called the *general theory of evolution.*

Thus when one uses the word "evolution," the intended meaning should be made clear.

Sometimes people use the word "evolution" to suggest changes that always lead from a simple to a more complex state. Modern evolutionists, however, usually avoid such a restricted usage of the term. Although biologists believe that evolution does sometimes lead to greater complexity, they believe that other times it leaves complexity unaffected and that still other times it reduces complexity.

Evolutionism is a philosophical worldview. It suggests that life as we see it today resulted from a complex series of events over a period of time measured in billions of years. *Naturalistic* or *materialistic evolutionism* is the view that evolutionary changes occurred without the intervention of a supernatural force, that ultimately the history of life will be understood solely on the basis of physical and chemical forces and interactions. The term "evolutionist" is often used to designate a person who espouses naturalistic evolutionism; however, even creationists who believe in microevolutionary processes are, strictly speaking, "evolutionists."

Theistic evolutionism posits that life has experienced the history described by naturalistic evolutionists. Instead of this process resulting from mindless natural forces, theistic evolutionists believe that God directed the process. By contrast, *deistic evolutionism* suggests that God created life in some simple, initial form and then withdrew from the scene. Since that initial event life has been on its own, evolving in response to the impersonal forces of nature.

CREATIONISM AND THE CREATIONIST MOVEMENT

It comes as a surprise for some people to learn that creationists disagree with one another more than evolutionists disagree with one another. Creationist disagreements have led to widely divergent views on everything from the age of the earth to the extent of biological change. Here I will summarize only the high points of this diversity of belief. Interested readers are encouraged to peruse Ronald L. Numbers' excellent history on the topic, *The Creationists* (1992).

One of Charles Darwin's prominent American disciples was Harvard botanist Asa Gray (1810-1888). Gray was an orthodox Christian believer; however, he found Darwin's arguments as applied to the nonhuman, natural world compelling, and as a result he developed a theistic evolutionary perspective. Not only had God created life in the beginning, he said, but he had directed its subsequent evolution over the ages. Moreover, Gray believed in a divinely ordained origin for humans.

Arnold Guyot (1807-1884), a New Jersey geologist and an active member of the Presbyterian church, gave more credence to the biblical creation story than did Gray. Like Gray, Guyot accepted long-age evolution, but Guyot believed that Genesis 1 provided a broad outline of Earth's history. Each "day" of the creation story represented a long period of geologic time. Also like Gray, Guyot found it impossible to accept the notion of human evolution. No amount of time would ever "suffice to make of the monkey a civilizable man," he opined. Guyot's *day-age theory* was popularized by two prominent geologists, James Dwight Dana (1813-1895) and John William Dawson (1820-1899).

Another approach, the so-called *gap theory,* was popular among more conservative creationists of the late nineteenth and early twentieth centuries. This was the notion that between the initial creation of "the heavens and the earth" mentioned in Genesis 1:1 and the start of creation week described in later verses, there was a blank spot in the biblical record. During this unreported time interval, much of the geological column was formed, including the record of life before creation week. The gap theory was given a great deal of support from C. I. Scofield (1843-1921), who wrote an annotation in his popular *Scofield Reference Bible* (1909) to the effect that Genesis 1:1 "refers to the dateless past, and gives scope for all the geological ages." Presbyterian minister and evangelist Harry Rimmer (1890-1952) was an outspoken proponent of this view.

Many people assume that the creationism popular among fundamentalists today was the most common form of creationism during the late nineteenth and early twentieth centuries. As pointed out by Numbers, however, "To find a creationist [during this period] who insisted on the recent appearance of all living things in six literal days, who doubted the evidence of progression in the fossil record, and who attributed geological significance to the biblical deluge, one has to look far beyond the mainstream of scientific thought" (*The Creationists,* p. 11). Although some practicing scientists and science educators of the late nineteenth century rejected the notion of evolution, most of these people tended to expand Earth's history far beyond six thousand years.

During the latter part of the nineteenth century, two brothers stand out as the most ardent defenders of what is today called conservative creationism. New York businessman Eleazar Lord (1788-1871), the older of the two, believed in a literal six-day creation about six thousand years ago. He was very critical of geologists, including those who wished to find compromises between scripture and long-age chronologies. According to Lord, Noah's flood was responsible for the fossil-bearing strata, and the progressive sequence of fossils contained in the geologic column was the result of sequential burial by the rising flood waters. His younger brother David Nevins Lord (1792-1880) entertained similar views but differed somewhat in his interpretation of the impact of the flood. According to David, significant portions of the geologic column were produced during upheavals that occurred between creation and the flood as well as since the abatement of the flood waters.

The single most influential conservative creationist of the early twentieth century was Seventh-day Adventist educator George McCready Price (1870-1963). Besides two years of college, he had spent one year at a teacher-training school where he took some elementary classwork in the natural sciences, including mineralogy. With these meager scientific credentials he proceeded to write more than twenty books and hundreds of magazine articles in support of conservative creationism. He also corresponded with a number of prominent evolutionary scientists of his day, including paleontologist David Starr Jordon (1851-1931), the president of Stanford University.

Never one to mince words, Price described Charles Darwin as "of the slow, unimaginative type so frequently found among English country squires [and] . . . singularly incapable of dealing with the broader aspects of any scientific or philosophic problem" (Price, 1934, p. 118). Darwin and other "field naturalists" were "mere children," he wrote, "when attempting to handle the larger problems of science" (Numbers, 1979).

According to Price, the theory of evolution was wholly dependent on the "geological ages hoax." He believed that if he could destroy the basis upon which geologists argued for long ages, he could knock the props out from beneath evolutionism. He was particularly critical of three assertions made by geologists: (1) that there was predictable order to the fossils in the geologic column; (2) that overthrusts, slippages of older rock strata over younger strata, sometimes occurred; and (3) that an ice age had been responsible for covering large areas of the northern hemisphere with glaciers.

Like his predecessor Eleazar Lord, Price argued that the Genesis flood had laid down nearly all the rock strata, although Price opposed any notion of a predictable fossil sequence. To him the fossil sequence was nothing more than an artificial contrivance created by geologists to provide underpinning for evolutionary theory. The flood had thoroughly jumbled the evidence of life past. No such thing as overthrusts existed—they were merely convenient explanations by geologists to explain out of sequence fossils. As for the ice age, all the supposed evidence for it could easily be attributed to the effects of the flood.

Price despised field work and never took an opportunity to examine at first hand most of the evidence he discussed. His effectiveness as a writer convinced many of his lay readers of his interpretations, but his views eventually ran into trouble when one of his former students and fellow Adventist, Harold W. Clark (1891-1986), visited the oil fields of Oklahoma and Texas in 1938. Clark, who had been a loyal disciple of Price, was shocked to discover that a predictable fossil sequence was the basis for the success of the entire petroleum exploration industry. Moreover, further study and travel convinced Clark that overthrusts were a geological fact of life and that widescale glaciation had occurred over much of the northern hemisphere.

Clark wrote his own geology text, *The New Diluvialism* (1946), in which he outlined his more progressive views. Still a firm believer in a universal flood and a recent creation, Clark posited that rising floodwaters had wiped out successive ecological zones, leading to the predictable sequence of fossils we see today. During pre-flood times, however, life zones were different from what they are today. For example, small oceans occurred at many elevations, not just at what we call sea level in the modern world. Clark's *ecological zonation theory,* as it was called, was very appealing to creationists because it allowed them to accept the validity of the geologic column but at the same time reject the long time periods of contemporary geologists. Clark's model continues to receive support from some quarters of the creationist community.

Meanwhile, the results of radiometric dating and other discoveries seemed to confirm the assertion of geologists that both the earth and life are very old. Many Christians found it impossible to continue to deny the evidence for long geological ages. One such person, Baptist theologian Bernard L. Ramm (1916-1993), published a widely discussed book in 1954 titled *The Christian View of Science and Scripture.* While acknowledging the influence of Price as "staggering," Ramm rejected the view that faith in scripture necessitated belief in a recent creation and a worldwide flood. Instead, he favored progressive creationism, a belief that God chose to create organisms of different types at widely different times. According to Ramm, "creation was *revealed* in six days, not *performed* in six days."

More conservative elements within evangelicalism were less than enthusiastic about Ramm's efforts, however, feeling that he had sold out to "modernism." How could anyone trust scripture on issues such as the virgin birth and the resurrection of Jesus, they asked, if it could not be trusted on the issue of creation? Thus in 1961 John C. Whitcomb, Jr. (b. 1924), a fundamentalist professor of Old Testament, and Henry M. Morris (b. 1918), a Baptist hydrologic engineer, published *The Genesis*

Flood, in essence a rebuttal to Ramm's book. Whitcomb and Morris hoped to convince readers that the great flood described in Genesis was a historic event of worldwide proportions. Acceptance of their view, they believed, would foster confidence in the literal interpretation of other parts of the Bible as well, particularly the creation account. To Whitcomb and Morris, the theory of evolution, as well as accommodationist views like Ramm's, directly contradicted the plain words of scripture and constituted one of the most serious threats to Christianity. Their message struck a responsive chord in the hearts of fundamentalist Christians throughout the world.

The Genesis Flood incorporated many of George McCready Price's views on geology, even some of those questioned by Clark. Unlike Price, though, Whitcomb and Morris were well credentialed, each with an earned doctoral degree. Moreover, they represented a broader spectrum of American Protestantism than Price, whose Seventh-day Adventism made him suspect to many people from mainline churches. Thus, Whitcomb and Morris were taken more seriously than Price, even though their perspectives differed little from his. The response to Whitcomb and Morris' book was overwhelming and far reaching. Creationist institutes were founded, creationist magazines and books were published, and creationist conventions were held. During the 1980's and 1990's, creationism played a significant role in the fundamentalist revival sweeping America.

The appearance of Whitcomb and Morris' book coincided with a post-Sputnik initiative to improve science education, including the teaching of evolution, in America's public school systems. Thus, just as conservative American Christians were converting to Whitcomb and Morris' perspectives on Earth's history, the principles of evolutionary biology and geology were receiving greater emphasis in America's classrooms. The resultant conflict led to the introduction of legislation mandating the teaching of "creation science" in the public schools of several states. The ensuing legal battles, particularly those in California, Arkansas, and Louisiana, received widespread media coverage. In the long run, creationists were unable to achieve many of their goals through the legal system, but their efforts paid off in terms of publicity received. By the mid-1990's, creationists had begun to focus their efforts on the control of local school boards rather than over state educational systems and, in the process, were achieving better success.

Just as with terms used by evolutionists, creationist terms can be confusing and misleading. "Creation" is generally understood to mean the instantaneous origin of objects in the universe through action of supernatural power or will. Many supporters of creation believe that the Hebrew scriptures provide a clear description of the creator's attributes—that, among other things, he is omnipotent (all-powerful), omniscient (all-knowing), and omnipresent (present everywhere). Supporters of creation do not necessarily believe in the Judeo-Christian God, however, and some even prefer to think of the creator in rather nebulous terms.

Strictly speaking, "creationism" is a philosophical perspective that presupposes the existence of a supernatural creator. Currently, however, this term usually denotes a sociopolitical movement of conservative, right-wing Christians who occupy a narrow range of the creationist belief spectrum. Creationism in this restricted sense usually presupposes the following tenets:

1. As the infallible, authoritative word of God, the Bible provides the only historically accurate record of the creation, Noah's flood, and other early events.
2. Creation occurred in six literal days about six to ten thousand years ago.
3. Noah's flood covered the entire world and was responsible for most of the geologic column and its accompanying fossil record.
4. Only minor alterations in living things have occurred over time. All the main types of organisms, including humans, were created in much the same form as we see them today (Morris, 1993).

In this view a "creationist" is a person who accepts these basic tenets, but many people who accept broader interpretations of creation are essentially creationists as well.

Recently, much has been made of the term "scientific creationism" or "creation science," the view that creationism provides scientifically testable hypotheses about the history of the earth. According to scientific creationists, all scientific data and the events described in scripture are in perfect agreement. When science and scripture seem to clash, either we misunderstand scripture or we misinterpret science. In actual practice, because scientific creationists believe that scripture, unlike science, is simple to read and easy to understand, it is science that they usually feel must be reinterpreted to fit the obvious meaning of scripture.

"Flood geology," the brainchild of George McCready Price in its twentieth century incarnation and a close ally of scientific creationism, provides scientific creationists with their basis for a belief in a young earth chronology. Flood geology is an extreme form of *catastrophism*, the notion that catastrophes, more than slow, gradual processes, have been responsible for the major features of the geologic column. Although evolutionary geologists acknowledge the occurrence of numerous catastrophes in their models of Earth's history, flood geologists place most of their emphasis on one such event, Noah's flood. They assert that this single, year-long deluge was responsible for significant portions, if not most, of the geologic column. The reason for this assertion is that it is the only event mentioned in scripture which seems to have been sufficiently powerful and all-inclusive to produce the earth's massive sedimentary rock layers in the short period of time available in a biblically based chronology.

During the 1990's, a more sophisticated theory of creationism became popular, the "theory of intelligent design." While many aspects of this theory are similar to the views of William Paley, proponents such as Michael Behe (1996) typically argue from a biochemical viewpoint. Intelligent design theorists believe that certain complex features of organisms could not have evolved gradually, but must have appeared suddenly as complete functional units. Some design theorists avoid identifying their postulated designer as the Judeo-Christian God, at least overtly. Also, many design theorists reject young earth creationism, avoid classification as creationists, and allow for significant amounts of evolutionary change.

CONCLUSION

The broad scientific community has responded to creationist initiatives with alarm. Most scientists view creationism as a form of pseudoscience and a threat to the integrity of the scientific enterprise. Consequently, large numbers of books, magazine articles, films, and other resources have been produced to head off the creationist offensive. Creationists have responded with their own flood of materials. One outcome from all of this is that scientists now spend more time than they did before thinking about the nature of science, its strengths and limitations, and how best to communicate scientific principles to the general public. The creationist movement has also stimulated more liberal Christians and adherents of other religious perspectives to consider more seriously the interaction between faith and science in the modern world.

Many evolutionists would like to see creationism eradicated from the arena of public thought, while many creationists would like to see creationism triumph over evolutionism. Neither of these outcomes, however, seems very likely in the foreseeable future. Both viewpoints are too deeply ingrained in mainstream culture.

REFERENCES

Behe, Michael J. *Darwin's Black Box: The Biochemical Challenge to Evolution.* New York: The Free Press, 1996.

Clark, Harold W. *The New Diluvialism.* Angwin, CA: Science Publications, 1946.

Darwin, Charles. *On the Origin of Species by Means of Natural Selection, or the Preservation of Favoured Races in the Struggle for Life.* London: J. Murray, 1859.

Desmond, Adrian, and James Moore. *Darwin.* New York: Warner Books, 1991.

Gould, Stephen Jay. *Ever Since Darwin.* New York: Norton, 1977.

Morris, Henry M. *History of Modern Creationism.* Second edition. Santee, CA: Institute for Creation Research, 1993.

Numbers, Ronald L. *The Creationists.* New York: Knopf, 1992.

Numbers, Ronald L. " 'Sciences of Satanic Origin': Adventist Attitudes Toward Evolutionary Biology and Geology," *Spectrum* 9 (January 1979), pp. 17-30.

Paley, William. *Natural Theology: Or, Evidences of the Existence and Attributes of the Deity, Collected from the Appearances of Nature.* London: Fauldner, 1802.

Price, George McCready. *Modern Discoveries Which Help Us to Believe.* New York: Revell, 1934.

Price, Peter W. *Biological Evolution.* Fort Worth, TX: Saunders College Publishing, 1996.

Ramm, Bernard. *The Christian View of Science and Scripture.* Grand Rapids, MI: Eerdmans, 1954.

Scofield, Cyrus I. (ed.). *The Scofield Reference Bible.* New York: Oxford University Press, 1909.

Whitcomb, John C., Jr., and Henry M. Morris. *The Genesis Flood.* Philadelphia, PA: Presbyterian and Reformed, 1961.

Chapter 2

HISTORICAL REFERENCES

The fifty-four references listed in this chapter include influential books published before 1981. These served as important antecedents to modern evolutionary and creationist views. Background information on these books and their authors was obtained from the following sources referenced in Chapters 3 and 7: Desmond and Moore (1991), Futuyma (1986), Gould (1987, 1993), Mayr (1982), Numbers (1992), and Price (1996). Unlike the alphabetical listings in other chapters, works listed here are arranged chronologically by date of original publication. If the cited work is not a first edition, year of first publication is indicated in brackets. Total numbers of pages are provided only for single-volume works.

Copernicus, Nicolaus. *On the Revolutions of the Heavenly Spheres.* Volume 16 of Great Books of the Western World, Robert Maynard Hutchins, editor-in-chief. Chicago, IL: Encyclopedia Britannica, [1543] 1952; pp. 497-838.

Copernicus (1473-1543) of Poland was trained as a mathematician, astronomer, canon lawyer, and physician. The prevailing view of his time was that the sun moved around the stationary earth, a notion backed up by intricate mathematical calculations by Ptolemy during the second century A.D. and coopted by the Christian Church. Copernicus, however, believed that the physical evidence pointed toward a heliocentric system, with the earth revolving around the sun. *On the Revolutions* is dedicated to Pope Paul III, with Copernicus judiciously telling the pope that among mathematicians, at least, his astronomical "labours will be seen to contribute something to the ecclesiastical commonwealth." Copernicus begins his book by describing "all the locations of the spheres or orbital circles together with the movements which I attribute to the earth"—in other words "the general set-up of the universe." He, then, correlates "all the movements of the other planets and their spheres or orbital circles with the mobility of the Earth." The book is chock full of data and calculations. Copernicus provided us with our modern view of the universe. He died just as the book was being published. Seventy-two years later, *On the Revolutions* was banned as heretical by the Congregation of the Index.

The Holy Bible. King James version. Philadelphia, PA: Lippincott, [1611] 1842; 823 pp.
The Holy Bible is a collection of sixty-six books written over a period of several hundred years. The Old Testament is revered by the Jewish people, whereas both the Old and New Testaments are esteemed by Christians. Many biblical books make reference to the creation, most notably Genesis, Job, Psalms, and John. The modern creation/evolution controversy often centers around interpretations of Genesis 1-11. These chapters fill a dozen or so printed pages and contain two creation narratives, two genealogies, and the account of Noah's flood. Until recently, many versions of *The Holy Bible*, including the example cited here, provided in a marginal note the year "4004 B.C." as the date of creation. This date had been calculated by Archbishop James Ussher on the basis of genealogical data, king lists, and other historical data, and was taken to be inspired by some Christians. Genesis 1 consists of a very organized account of creation—six days of creative activity followed by a day of rest; here the Hebrew name for God is Elohim, and Elohim is pictured as majestic and transcendent. By contrast, Genesis 2 contains a more loosely arranged creation narrative, with a somewhat different ordering of events; here the Hebrew name for God is Yahweh, and Yahweh is pictured as personable and approachable. The story of Noah's flood appears in Genesis 6-9, whereas the genealogies for the ten pre-flood and ten post-flood patriarchs are found in Genesis 5 and 11. Scientific creationists generally assume historical and scientific accuracy for Genesis 1-11, whereas more liberal interpreters allow for varying levels of symbolic interpretation.

Bacon, Francis. *Novum Organum.* New York: Collier & Son, [1620] 1902; 290 pp.
Bacon (1561-1626), Baron of Verulam and Viscount St. Alban, was a statesman, courtier, and philosopher. Principles enunciated in *Novum Organum* have been important in the development of scientific meth-

odology. Bacon deplores the sorry state of human understanding of the world and sets forth a new means for achieving knowledge. He posits that there are two methods of gaining knowledge. In the first, *deductive* method, facts about nature are ascertained, and from these facts broad generalizations about nature are made. From these broad generalizations, propositions about intermediate particulars are deduced. In the second, *inductive* method, data from the natural world are used to form low-level axioms which, in turn, are used to develop intermediate axioms, and so on. Ultimately broad generalizations about nature are achieved. According to Bacon, only the second, inductive, method is appropriate as a means to gain scientific knowledge. Like Bacon, creation scientists often define science as a body of inductively ascertained factual knowledge about the world. In this view, evolutionary theory lies outside the realm of science. Most members of the scientific community, however, reject a strictly Baconian approach to achieving knowledge and view the scientific method as incorporating both deductive and inductive processes.

Galilei, Galileo. *Dialogue Concerning the Two Chief World Systems—Ptolemaic & Copernican.* Translated by Stillman Drake; foreword by Albert Einstein. Second edition. Berkeley: University of California Press, [1632] 1967; 496 pp.

When he wrote *Dialogue*, Galileo (1564-1642) was serving as mathematician and philosopher to the Grand Duke at Florence. He had previously taught mathematics at Pisa and at Padua. Galileo was convinced that the Copernican cosmology was correct—that the earth moves around the sun. By contrast, Ptolemy had taught that the sun moves around the earth, and the Roman Church had incorporated this geocentric view into its teachings. In *Dialogue*, Galileo sets out to demonstrate the logic of the Copernican, heliocentric system. He does this in the form of a fictional conversation that takes place over a period of four days among three friends: Saliati, Sagredo, and Simplicio. Using this conversation as a pedagogical vehicle, Galileo accomplishes three things: first, he shows that experiments done on earth are inadequate measures of its mobility; second, he examines "celestial phenomena," which provide irrefutable evidence favoring the Copernican view; and third, he shows that the ocean tides can be explained within the context of a moving earth. For his efforts, Galileo was summoned to Rome where he was tried by the Inquisition for his Copernican views and for meddling in "high matters"—the theological implications of his scientific discoveries. He spent the rest of his life under house arrest for his heretical views.

Lightfoot, John. *A Few, and New Observations, Upon the Book of Genesis. The Most of Them Certain, the Rest Probable, All Harmless, Strange, and Rarely Heard of Before.* London: T. Badger, 1642; 20 pp.
Lightfoot (1602-1675) was a biblical scholar and a vice chancellor of Cambridge University. In this pamphlet he states that "heaven and earth, centre and circumference, were created together, in the same instant, and clouds full of water." Moreover, "man was created by the Trinity on the twenty-third of October, 4004 B.C., at nine o'clock in the morning." Lightfoot utilized data from the genealogies of Genesis 5 and 11, and the assumption that creation would have occurred on the autumnal equinox, to arrive at his date.

Ussher, James. *The Annals of the World. Deduced from the Origin of Time, and Continued to the Beginning of the Emperour Vespasian's Reign, and the Totall Destruction and Abolition of the Temple and Common-wealth of the Jews.* London: Tyler, 1658; 907 pp.
Ussher (1581-1656) was archbishop of Armagh and head of the Anglo-Irish church. He was a language scholar, and he had taught and administered at Trinity College, Dublin, before assuming his ecclesiastical post. Ussher begins his massive tome with the words: "In the beginning God created Heaven and Earth, Gen. 1. v. 1. Which beginning of Time, according to our Chronologie, fell upon the entrance of the night preceding the twenty third day of Octob. in the year of the Julian Calender, 710"—meaning, as Ussher notes in the margin, 4004 B.C. as we measure time today. "Upon this first day therefore of the world, or Octob. 23. being our Sunday, God, together with the highest Heaven, created the Angels. Then having finished, as it were, the roofe of this building, he fell in hand with the foundation of this wonderful Fabrick of the World, he fashioned this lowermost Globe, consisting of the Deep, and of the Earth." Ussher's year of 4004 B.C. represents an impressive amount of scholarship and is similar to creation dates calculated by other seventeenth century academicians. It is based on the genealogies of Genesis 5 and 11, the years of the kings of Israel and Judah, and extra-biblical dates of events in Babylon, Persia, and the Roman empire. The specific date of October 23 is somewhat arbitrary but reflects certain eschatological

and anti-Roman Catholic sentiments of Ussher. His 4004 B.C. date for creation was included in the margins of some printings of the King James version of the Bible until the last half of the twentieth century.

Burnet, Thomas. *The Sacred Theory of the Earth.* Carbondale, IL: Southern Illinois University Press, [1690-1691] 1965; 412 pp.
Burnet (1635?-1715) was an Anglican clergyman who first published this book in Latin as *Telluris theoria sacra.* Burnet's purpose is primarily theological, not geological—to show that the present world is not the perfect planet that God created at the beginning. Due to the effects of sin and the devastating action of the flood, it is a ruined world which will ultimately be recreated. *The Sacred Theory of the Earth* consists of four books. Book I addresses Noah's flood. Burnet believed that the original earth had a smooth, regular surface. This surface rested on a subterranean sea which, at the time of the flood, burst forth through great fissures. This action caused broken pieces of the crust to tip up on their sides to form mountain ranges. Book II describes the earth before the flood. At this time the earth was not tilted on its axis. The poles were slightly elevated relative to the equator. Thus, rivers flowed from the poles to the equatorial tropics where the water evaporated and returned, once again, to the poles. Book III argues that the world will once again be brought to a state of chaos, this time by fires lit by volcanoes. The soot from the fires will settle into concentric, density-dependent spheres, and the earth will once again be perfect. Book IV describes the new heavens and the new earth that will appear following the conflagration of Book III. The world will then be a bright star, which Burnet believes was its original state. He contends that this return "to the same state again, in a great circle of Time, seems to be according to the methods of Providence; which loves to recover what was lost or decay'd." And "There we leave it," writes Burnet, "Having conducted it [the earth] for the space of Seven Thousand Years, through various changes from a *dark Chaos to a bright Star.*"

Ray, John. *The Wisdom of God Manifested in the Works of the Creation.* Eleventh edition. London: Innys, [1691] 1743; 405 pp.
John Ray (1627-1705) was a fellow of the Royal Society of London. His *Historia Plantarum* described over eighteen thousand plant species and is considered to be an important antecedent to our modern clas-

sification system. *The Wisdom of God* is both an argument from design and a compendium of astute natural history observations. Ray's first purpose for the book is to establish belief in God, "a Matter of the highest Concernment to be firmly settled and establish'd"; second, he seeks to "illustrate some of his [God's] principal attributes"; and third, he hopes "stir up and increase in us the Affections and Habits of Admiration, Humility, and Gratitude." He sets out to accomplish his purposes through an examination of the "Works of Creation," meaning "the Works created by God at first, and by him conserv'd to this Day in the same State and Condition in which they were first made." Ray rejects spontaneous generation, Aristotle's view that the world is co-eternal with God; the Epicurean notion that the world was made by a casual concurrence and cohesion of atoms; and the Cartesian hypothesis that God created matter, put it into motion, and by natural law it became the world. Ray provides evidence of God's designership by examining water, tides, fire, air, meteors, plants, animals, and humans. Of the "body of Man," writes Ray, there is "Nothing deficient or redundant."

Woodward, John. *An Essay Toward a Natural History of the Earth and Terrestrial Bodies Especially Minerals, as also of the Sea, Rivers, and Springs, with an Account of the Universal Deluge, and of the Effects that It Had Upon the Earth.* London: Wilkin, 1695; 277 pp.

Woodward (1665-1728) was a professor of medicine at Gresham College and a fellow of the Royal Society. In this book, Woodward interprets the history of the earth in terms of Noah's flood, an event designed to provide a habitable world for fallen humanity. He writes "that the whole Terrestrial Globe was taken all to pieces and dissolved at the Deluge," and that our present world "was formed out of that promiscuous Mass of Sand, Earth, Shells." But, Woodward notes, this "promiscuous Mass" is not a random jumble of sediments: "Marine Bodies are now found lodged in those Strata according to the Order of their Gravity," with the heaviest structures lying deepest and the lightest structures lying closest to the surface. Woodward insists on the universality of the flood. Much of the water, he says, came from the "Great Abyss," a huge "Subterranean Reservatory" beneath the earth's crust. Unlike some of his contemporaries, he believes that marine fossils are the remains of once living animals. The flood was responsible for placing them on what is now land. Woodward also discusses the opinions of other writers on the topic, the formation of minerals and metals, the nature of the pre-flood earth, and alterations of the earth since the time of the flood.

Whiston, William. *A New Theory of the Earth, from Its Original, to the Consummation of all Things. Wherein the Creation of the World in Six Days, the Universal Deluge, and the General Conflagration, as Laid Down in the Holy Scriptures, Are Shewn to Be Perfectly Agreeable to Reason and Philosophy.* New York: Arno, [1698] 1978; 388 pp.

Whiston (1667-1752) was chaplain to the Lord Bishop of Norwich and a fellow of Clare-Hall, Cambridge. His purpose in *A New Theory of the Earth* is "to account for the Creation of the World, agreeable to the description thereof in the Book of Genesis." His central proposition is that "The Mosaick Creation is not a Nice and Philosophical account of the Origin of All Things; but an Historical and True Representation of the Formation of our single Earth out of a confused Chaos, and of the successive and visible changes thereof each day, till it became the habitation of Mankind." He argues that Genesis 1 deals only with the earth itself, not the entire universe. The pre-earth chaos that formed the earth, he proposes, "was the Atmosphere of a Comet." He also states numerous other propositions, including that the mountains are high because they are less dense than other parts of the earth's surface; that "the Annual Motion of the Earth commenc'd at the beginning of the Mosaick Creation; yet its Diurnal Rotation did not till after the Fall of Man"; that the Garden of Eden occurred "on the Southern Regions of Mesopotamia, between Arabia and Persia"; that the "Original Orbits of the Planets, and particularly of the Earth, before the Deluge, were perfect Circles"; that a comet passing close to the earth during the flood gave the earth its elliptical orbit, as well as its tilted axis; that the habitable earth is situated on the surface of "a deep and vast Subterraneous fluid"; that "The Temper of the Humane Body was more soft, pliable, and alterable than now it is"; that "The Antediluvian Air had not large, gross Masses of Vapours, or Clouds, hanging for long seasons," and no "great round drops of Rain"; that some of the flood waters were of "celestial" origin, whereas some were from the waters beneath the earth; and that the present-day geological strata dropped out of the flood waters "according to the Law of Specifick Gravity." Each of these propositions, along with his many other hypotheses, is discussed in detail.

Buffon, Georges Louis. *Histoire naturelle, général et particulière.* Forty-four volumes. Paris: Imprimerie Royale, puis Plassan, 1749-1804.
Buffon (1707-1788) was one of the foremost scientists of the eighteenth century. Thirty-five of the volumes of this massive work were published before Buffon's death, while the remaining nine volumes appeared posthumously. Virtually every aspect of natural history is addressed, from minerals to humans. Over the long course of writing *Histoire naturelle*, Buffon seems to have changed his perspectives. He writes from a seemingly atheistic perspective in his first three volumes, then later endorses a more deistic approach—the view that the world is governed by divinely created natural law. Buffon notes that evolution by common descent is a possibility but provides three reasons to reject this view: (1) no one has ever seen the production of a new species; (2) change from one species to another is prohibited by hybrid infertility; and (3) intermediate forms, such as would be predicted on the basis of gradual change, are not found. He concludes that while "it cannot be demonstrated that the production of a species by degeneration from another species is an impossibility for nature, the number of probabilities against it is so enormous that even on philosophical grounds one can scarcely have any doubt upon the point." Spontaneous generation, he writes, was the source of the first individual of every species. Moreover, the number of possible species is limited only by the number of ways in which organic molecules can be arranged. Variation within species occurs as the result of environmental, not hereditary, influences. While Buffon rejected the concept of evolution, Europeans became aware of evolutionary ideas through his writings.

Linne, Carl von. *Species Plantarum.* Two volumes. Holmiae, Sweden: L. Salvii, 1753.
Linne (1707-1778), better known as Linnaeus, developed the binomial system of nomenclature, the practice used today to designate each species with a unique, two-part name—its scientific name. Prior to Linnaeus' invention, long and complicated Latin phrases were used to refer to specific types of organisms, an inefficient and tedious method. Linnaeus believed that he had been divinely chosen to classify the plants and animals God had created. Apparently he was successful in convincing other people of this belief as well, for he was held in great awe by his contemporaries. *Species Plantarum* is the first work to use the binomial nomenclature system, where it is

applied to plant species. Based on their commonalities, Linnaeus also groups the various species into larger, more inclusive categories—genera, orders, and classes. The same system is used in his later work, *Systema Naturæ* (1759), in which both plants and animals are listed. While Linnaeus viewed species as relatively fixed entities from the hands of the creator, his classification system provided later naturalists with evidence for patterned relationships among organisms, evidence that led to evolutionary theories of common ancestry.

Hume, David. *Dialogues Concerning Natural Religion.* Edited with Introduction by Henry D. Aiken. New York: Hafner, [1779] 1948; 95 pp.
Hume (1711-1776) was a Scottish philosopher who was deeply critical of the belief in miracles, immortality, the religious basis of morality, and religion in general. He was, in short, skeptical of any view pertaining to God. Nonetheless, he maintained a lifelong interest in the phenomenon of religion, if only as an object of critique. *Dialogues Concerning Natural Religion* is written in the form of a conversation among Cicero, Demea, and Philo, and narrated by Pamphilus. The central question addressed is the nature of God, which is determined to be an insoluble riddle. Of particular interest to participants in the creation/evolution controversy is the discussion of the argument from design. The defects of this argument are pointed out by Philo, who notes that before one can posit a designer, one must establish the existence of a worldwide pattern of design; this has not been established. Also, the argument from design argues on the basis of cause/effect relationships *within* nature, whereas a designer would affect nature from *outside*, seriously damaging the argument. Moreover, what is to say the universe is not simply the result of generation, just like a plant generates new plants without the action of a deity? And finally, even if we accept the notion of a designer, what caused the designer? Because that question is unanswerable, the entire question of the existence of a designer is devoid of meaning. William Paley's book *Natural Theology* was written in a effort to counter arguments like those in Hume's *Dialogues.*

Hutton, James. *Theory of the Earth with Proofs and Illustrations.* Four volumes. Edinburgh: Cadell, Junior, and Davies, 1795.
Hutton (1726-1797), a Scottish thinker and geologist, is the person

generally credited with the discovery of "deep time." Concepts presented in his verbose, nearly impenetrable *Theory of the Earth* were made accessible by his friend, John Playfair, who introduced them more effectively in *Illustrations of the Huttonian Theory of the Earth* (1802). Hutton views the world as an endlessly cycling, Newtonian machine. Each cycle consists of three stages: first, erosional forces break apart rocks; second, the sediments formed by erosion settle into horizontal strata; and third, heat from the weight of the sediments melts the lower strata, and movements of the resultant magma cause extensive uplift. Thus, continents form where oceans used to be. The uplifted continents are now, in turn, subject to erosion—and the cycle repeats. Hutton believes this cyclicity serves the purposes of humans. Had we not found evidence of this cyclicity, he writes, "we should have reason to conclude, that the system of this earth has either been intentionally made imperfect, or has not been the work of infinite power and wisdom." Hutton views fossils as representing organisms still present on the earth; no new species appear, and no species go extinct. The one exception may be humans, which he allows may be recent additions to the earth. To Hutton, the rocks of the earth provide "no vestige of a beginning,—no prospect of an end."

Darwin, Erasmus. *Zoonomia, or, The Laws of Organic Life.* Volume 1. New York: Swords, 1796; 433 pp.
Erasmus Darwin (1731-1802) was a prominent English physician, poet, and free-thinker, as well as the grandfather of Charles Darwin. The purpose of *Zoonomia*, he writes in the preface, "is an endeavor to reduce the facts belonging to Animal Life into classes, orders, genera, and species; and by comparing them with each other, to unravel the theory of diseases." Such a theory, he writes, "should bind together the scattered facts of medical knowledge, and converge into one point of view the laws of organic life, [and] would thus on many accounts contribute to the interest of society." Darwin believes that nature consists of two essences: spirit, which produces motion, and matter, which receives and communicates motion. Much of the book is a commentary on the diseases of various organ systems in light of these essences. Darwin takes a comparative approach, believing "that all warm-blooded animals" are related to one another, having all "arisen from one living filament." Along with many of his contemporaries, he believed that life contained an innate drive to improve. Charles Darwin read his grandfather's book, but the grandson's views

on evolution developed along decidedly different, materialistic lines from those of the elder Darwin.

Laplace, Pierre Simon. *The System of the World.* Two volumes. London: Richard Phillips, [1796] 1809; 754 pp.
Laplace (1749-1827), a prominent French scientist, was interested in how the solar system had formed. In *The System of the World*, he rejects the notion that it arose by pure chance, for it exhibits too much apparent order. Laplace proposes that the planets were formed from the atmosphere of the sun. This atmosphere, said Laplace, once extended beyond the orbit of the farthest planet; as the sun cooled, the atmosphere condensed into a series of Saturn-like rings. The rings were aligned along a plane consistent with the sun's equator. The rings coalesced, and as a result, the planets were formed. Moons developed in much the same way, from planetary rings that coalesced. These processes occurred, believed Laplace, as an outworking of natural law. Laplace's "nebular hypothesis," as it came to be known, was a naturalistic model, one that did not invoke the direct involvement of a deity. In this respect, Laplace's views were consistent with those of a growing number of late eighteenth and early nineteenth century scientists, and they helped to pave the way for the Darwinian revolution. While the nebular hypothesis became the generally accepted view of solar system origins by the mid-nineteenth century, evidence contrary to this view accumulated, so that by the end of that century it had been rejected.

Malthus, Thomas R. *An Essay on the Principle of Population.* Sixth edition. London: Reeves and Turner, [1798] 1888; 551 pp.
Malthus (1766-1834) was a fellow at Jesus College, Cambridge, and a professor of history and political economy at East India College, Hertfordshire. Malthus was concerned with how increases in the human population heighten human poverty and misery. The first version of this essay was published in 1798 and was subsequently expanded and updated. Both Charles Darwin and Alfred Russel Wallace were greatly influenced by Malthus' *Essay* as they independently developed their respective versions of the theory of natural selection. Malthus begins by examining checks on the populations of various human populations in the past. He then discusses various means proposed to deal with the evils arising from overpopulation: systems of equality, emigration, poor laws, agricultural and commercial systems,

and corn laws. He proposes that humans must control their sexual passions in an effort to restrain population growth, and he reviews the dire prospects for humans if they fail to do this. He contends that the human tendency to provide charity to the underprivileged masses simply exacerbates the population problem.

Paley, William. *Natural Theology; Or, Evidences of the Existence and Attributes of the Deity, Collected from the Appearances of Nature.* London: R. Faulder, 1802; 548 pp.

Paley (1743-1805) was an Anglican churchman, Archdeacon of Carlisle, and a fellow of Cambridge University. Arguments from design for the existence of God had been posited for centuries, but Paley gave these arguments particular force in *Natural Theology.* He begins by comparing his hypothetical reactions to finding a stone and a watch in a field. He says that he might be justified in assuming the stone "had lain there for ever." But he could not assume the same thing for the watch. The existence of a watch implies "that the watch must have had a maker . . . who comprehended its construction, and designed its use." This conclusion for the watch is warranted, he argues, because "its several parts are framed and put together for a purpose." He then suggests that organisms, like the watch, show even greater evidence of design. Much of the book is devoted to an application of this argument to the function of bones, muscles, blood vessels, instincts, plants, heavenly bodies, and numerous other natural objects. From this evidence he infers the existence of an "intelligent Creator" with certain attributes apparent from the objects of his creation. Paley's argument strongly influenced Darwin, who eventually substituted natural selection as the primary agent of design. Paley's argument for a creator is experiencing something of a resurgence today among "design theorists" who argue on the basis of the apparent "irreducible complexity" of some natural systems.

Lamarck, Jean-Baptiste. *Zoological Philosophy.* Chicago, IL: University of Chicago Press, [1809] 1984; 453 pp.

Lamarck (1744-1829), a prominent French zoologist, was one of the first trained scientists to develop a mechanistic theory of evolution. While his views have been ridiculed by modern biologists, he made important contributions outside of evolutionary theory, including publication of an important treatise on invertebrates. In *Zoological Philosophy* Lamarck states that a new theory is necessary to account for the graded series and great diversity of living things. He assumes that organisms developed gradually over long ages and that spontaneous generation is responsible for the emergence of microorganisms "when conditions are favorable." It is from these lower organisms that all other forms of life emerge. He postulates that, instead of going extinct, organisms change into other types of organisms over time. Lamarck suggests two causal forces in evolutionary change: an innate drive toward progress in organisms and the ability to respond to environmental change. Environmental change, he says, results in a change in the needs of an animal; the needs of an animal are satisfied by behavioral change; the behavioral change results in certain anatomical parts being used more than before; this greater use causes greater development of these structures; and this enhanced development is passed on to the animal's offspring. Thus, characteristics acquired during the lifetime of an animal are inherited by its young. Even though plants do not behave in the sense that animals do, "great alterations of environmental circumstances nonetheless lead to great differences in the development of their parts," differences that Lamarck says are passed on to their progeny as well.

Cuvier, Baron Georges. *Essay on the Theory of the Earth.* Fifth edition. Edinburgh: Blackwood, [1813] 1827; 550 pp.

Cuvier (1769-1832), secretary of the French Institute and professor and administrator of the Museum of Natural History, is considered to be the father of comparative vertebrate morphology. He was responsible for describing large numbers of vertebrate fossils. *Essay on the Theory of the Earth* is designed to "shew by what relations the history of the fossil bones of terrestrial animals connects itself with the theory of the earth." Just as humans have had their crises, says Cuvier, "Nature also has had her intestine wars, and . . . the surface of the globe has been broken up by revolutions and catastrophes." Cuvier believes that fossils are the remains of once-living organisms, not "sportings of nature," as some have believed. Moreover, he posits that the life-forms represented by fossils represent a succession of organisms through time, repeatedly destroyed by sudden and devastating catastrophes. Of these catastrophes, the most recent "cannot be referred to a much earlier period than five or six thousand years ago," and was no doubt the flood mentioned in Genesis. Following this event, "the small number of individuals dispersed by it have spread and propagated over the newly exposed lands" and human civilizations flourished. He mentions several theories of earth history and concludes that "It is evident, that, even while confined within the limits prescribed by the Book of Genesis, naturalists might still have a pretty wide range" of ideas." Much of the book provides descriptions of many of the vertebrate fossils studied by Cuvier.

Lyell, Charles. *Principles of Geology, Being an Attempt to Explain the Former Changes of the Earth's Surface by Reference to Causes Now in Operation.* Three volumes. London: Murray, 1830-1833.
Lyell (1797-1875) was an English lawyer and geologist. His *Principles of Geology* ranks as one of the great scientific works of the nineteenth century. Lyell argues that knowledge about the geologic past must be apprehended using a methodology that would eventually become known as *uniformitarianism.* He argues the principle of uniformity from four perspectives: uniformity of natural law, uniformity of process, uniformity of rate, and uniformity of state. In other words, historical geology must be studied from the viewpoint that the same natural laws have governed the same set of geologic processes from the beginning; furthermore, the rates of these processes have remained roughly the same over time, and the earth exists in a state of dynamic equilibrium. Widespread catastrophism is out of the question, he argues; only localized catastrophes are possible. Lyell also discusses the history of life in the context of geologic time. Species appear and disappear with clocklike regularity. Again, balance and uniformity are maintained. He believes that the appearance of organismal progress in the geologic column is an illusion created by an incomplete understanding of the fossil record. Humans are special and set apart from the rest of the animal world. Charles Darwin read Lyell's *Principles* on his *Beagle* voyage and was greatly influenced by this work. Later, Lyell and Darwin became friends, though Lyell resisted Darwin's transmutation theory. Toward the end of his life Lyell grudgingly gave in to some of Darwin's views but continued to reserve a special place for humans in the natural economy.

The Bridgewater Treatises on the Power, Wisdom and Goodness of God as Manifested in the Creation. Twelve volumes. London: William Pickering, 1833-1836.
This series of eight treatises in twelve volumes was funded by a bequest from the Earl of Bridgewater. It was written by prominent British scientists to demonstrate that every aspect of the natural

universe was purposefully designed by God, who is both wise and good. Rev. Thomas Chalmers and John Kidd examine the apparent adaptation of nature to the moral and intellectual constitution of humans; Rev. William Whewell shows how physics and astronomy declare God's glory; Sir Charles Bell discusses the intricacies of the human hand as evidence for a master designer; Peter Mark Roget develops his natural theology through an examination of plant and animal physiology; Rev. William Buckland interprets the rocks in terms of the gap theory; Rev. William Kirby details the behaviors and instincts of animals as evidence of a beneficent deity; and William Prout finds evidence for God in chemistry, meteorology, and digestive processes. Darwin read the treatise by Kirby during his voyage on the *Beagle*. These widely quoted treatises were extensions of Paley's argument from design. Despite their impact on the Christian community, *The Bridgewater Treatises* were scorned by free thinkers as preposterous.

Anonymous [Chambers, Robert]. *Vestiges of the Natural History of Creation. With a Sequel.* New York: Harper & Brothers, [1844] 1847?; 303 pp.
This volume was published anonymously, but eventually it was discovered to be the work of Robert Chambers (1802-1871). Chambers was a popular writer and editor of *Chambers' Encyclopedia. Vestiges* contains numerous factual errors and was harshly criticized by scientists when it appeared, but it was a smashing success with the public. While Chambers' views are very different from those of Darwin, he got Victorians thinking in evolutionary terms, thus paving the way for Darwin's *Origin of Species*. Chambers rejects special creation, believing instead that natural laws set up by the creator in the beginning were responsible for the origin and evolutionary development of the universe. According to Chambers, life emerged frequently and universally through spontaneous generation, although because conditions have changed, this process may no longer occur. On the basis of the fossil record, comparative anatomy, and comparative embryology, he argues for the slow and gradual evolutionary development of life. In his view, organisms repeat their past evolutionary stages as they develop embryonically: "Our brain goes through the various stages of a fish's, a reptile's, and a mammifer's brain, and finally becomes human." Moreover, opines Chambers, "after completing the animal transformations, it [the brain] passes through the characters in

which it appears in the Negro, Malay, American, and Mongolian nations, and finally is Caucasian." In short, *"The leading characters. . . of the various races of mankind are simply representations of particular stages in the development of the highest or Caucasian type"* (Chambers' emphasis). Thus Chambers, along with many of his contemporaries, was a committed racist.

Smith, John Pye. *The Relation Between the Holy Scriptures and Some Parts of Geological Science.* Fourth edition. Philadelphia: Peterson, 1850; 400 pp.
Smith (1774-1851) was a divinity tutor in the Protestant Dissenting College, Homerton. This volume consists of eight lectures presented at the Congregational Library. Smith believes that the "study of revealed religion . . . cannot but be in perfect harmony with all true science." He acknowledges, however, that geology seems to be opposed to a literal interpretation of the creation and flood stories. While he is committed to a high view of scripture, he is critical of what he sees as superficial interpretations of the Genesis account. "God created all things," he opines. "They were therefore not self-derived or eternal. But to pretend that there is any proof in Holy Writ, that God created them about six thousand years ago, and that to doubt this is infidelity, is to foist the received interpretation in the place of the inspired word, as well as to deal very harshly by our christian [sic] neighbour who thinks otherwise." Smith posits at least two creation events: one in the distant past, and a second, more localized, six-day creation in "the part of our world which God was adapting for the dwelling of man and the animals connected with him"—probably western Asia. This region had been laid waste before the beginning of this second, biblical creation. As for the question about death before sin, Smith opines that humans were exempt from death, but not animals. Smith considers the view that lions, wolves, and other carnivores were originally vegetarians as a "monstrous absurdity." He rejects the notion that Noah's flood covered the entire earth.

Lord, Eleazar. *The Epoch of Creation: The Scripture Doctrine Compared with the Geological Theory.* New York: Charles Scribner, 1851; 311 pp.
Eleazar Lord (1788-1871) was a conservative Presbyterian writer and businessman. He began studies at Princeton University but had to drop out due to poor eyesight. He became president of the Manhattan

Fire Insurance Company and later president of the New York and Erie Railroad. In *The Epoch of Creation*, Lord expresses his belief in a literal interpretation of Genesis 1 and holds that the earth is only about six thousand years old. He is highly critical of geologists and what he calls their "gratuitous assumptions." He rejects attempts by some Christian geologists to bend their interpretations of Genesis to geological theory, believing that geology should yield to a straightforward reading of scripture. Fossil-containing rocks, he believes, were formed by the Genesis flood; moreover, the rising floodwaters were responsible for the fossil sequence in the geologic column: "What can be more obvious," he writes, "than to suppose that the waters of the deluge rose, and progressively became charged with sedimentary matter, shell-fish should have been stifled and buried, before the placoids, ganoids, or other species of fish, fishes before reptiles, reptiles before birds, and birds before quadrupeds?" Thus, Lord anticipated the popular "ecological zonation theory" championed by Harold W. Clark ninety-five years later. Until the advent of George McCready Price's flood geology of the twentieth century, Lord was one of the last writers to defend a diluvial interpretation of the geologic column.

Lord, David N. *Geognosy; or the Facts and Principles of Geology Against Theories.* Second edition. New York: Franklin Knight, [1855] 1857; 412 pp.

David Lord (1792-1880) was a graduate of Yale University and a New York businessman. He and his older brother, Eleazar, were prominent conservative Christians during the nineteenth century. David edited the *Theological and Literary Journal* from 1848 to 1861, in which he frequently attacked geology and evolutionary biology. *Geognosy* focuses on the "false theories of geologists" and posits that "The whole Bible, as a revelation, thus stands or falls with the first chapter of Genesis." Lord repeatedly affirms his belief in a literal six-day creation which he believed happened "but six thousand years" ago. In his view, any theory that contradicts this interpretation of scripture is false. He believes that the geological strata were "derived from the interior of the earth" and deposited between creation and the flood, as well as after the flood. Lord downplays the impact of the flood on the geological strata, although his brother, Eleazar, believed that the fossil-bearing strata had their origin in the deluge. The Lord brothers were among only a few creationist writers of the late nineteenth century who insisted on a six-thousand-year-old earth.

Gosse, Philip Henry. *Omphalos; an Attempt to Untie the Geological Knot.* London: Van Voorst, 1857; 376 pp.
Gosse (1810-1888) was a biologist and creationist troubled by the growing evidence for an old earth and evolutionary change. *Omphalos* represents his attempt to solve this problem in the context of his biblical faith. He presumes that geologists are honest searchers for truth and that it would be wrong for them to overlook data from the natural world to satisfy perceived requirements of faith. "[T]hey cannot shut their eyes to the startling fact," Gosse writes, "that the records which *seem* legibly written on His created works do flatly contradict the statements which *seem* to be plainly expressed in His word." Gosses notes that previous writers "thought that the deluge of Noah would explain the stratifications, and [that] the antediluvian era accounts for the organic fossils," but that as "the 'Stone book' was further read, this mode of explanation appeared to many untenable." In its place Gosse posits "the Law of Prochronism." In this view, natural processes are cycles, and creation would have burst into each of these cycles at some point. Take the life cycle of a moth, he says—"pupa, larva, egg, moth, pupa, larva, &c. &c."—God would have made the first moth in one of these stages. Had the newly created moth been an adult, it would have looked many days older than the egg stage, whereas, in fact, it had just appeared. The moth's apparent, *prochronic* age, then, would differ from its real, *diachronic* age. Likewise, because of their prochronic ages, at the instant of creation trees had rings, Adam had an *omphalos* (Greek for "navel"), and the geologic strata had fossils. Much to his disappointment, most of Gosse's contemporaries rejected his views as absurd.

Miller, Hugh. *The Testimony of the Rocks, or, Geology in Its Bearings on the Two Theologies, Natural and Revealed.* Boston: Gould and Lincoln, 1857; 502 pp.
Tragically, only one day after the proofs for his preface were sent to the printer, Miller (1802-1856) shot himself in a fit of emotional illness. Earlier in life he had embraced the gap theory, the view that the biblical creation of six days was preceded by a "great chaotic gap" before which the geologic ages occurred. In the years just before writing this book, however, he had explored much more of the geologic column than before. This exploration led him to conclude that "No blank chaotic gap of death and darkness separated the creation to which man belongs" from the lives of other creatures. He came to the

conclusion "that the days of creation were not natural, but prophetic days, and stretched far back into the bygone eternity. . . . I have yielded to evidence which I found it impossible to resist." *The Testimony of the Rocks* consists of twelve lectures, most of which were delivered in public venues. Miller first describes the paleontological histories of plants and animals. He does not believe in any form of evolutionary change. Each species is a unique creation, with the sequence of creations constituting "the gradual fitting up of our earth as a place of habitation" for humans. He then compares the record of the rocks with the Genesis record of creation, and with both natural and revealed theology. He rejects the notion of a universal Noachian flood in favor of a local flood view. He criticizes the views of the "anti-geologists," who consider geologists to be "infidels." He concludes with two lectures on the lesser known fossil floras of Scotland.

Darwin, Charles. *On the Origin of Species by Means of Natural Selection.* London: Murray, 1859; 502 pp.
Darwin (1809-1882) was an English gentleman naturalist who, from 1831 to 1836, served as the unofficial naturalist on a round-the-world voyage on H.M.S. *Beagle*. He was particularly well suited to this task, for he was a keen observer and an energetic explorer. On the basis of his observations, he wrote four books on geology. His observations of plant and animal distributions led him to question the fixity of species concept, which he had earlier embraced. In its place, he worked out a theory of change by a process he called natural selection. In this view, organisms exhibit heritable variation. In a given environment, some variants reproduce more successfully than others—they are "selected." Over time, the traits of these successful organisms will become more pronounced. Over many generations accumulations of changes resulting from natural selection lead to the formation of new species. Darwin had ruminated on this theory for many years when, in 1858, he received a manuscript from a much younger English naturalist, Alfred Russel Wallace. Wallace had come up with a theory similar to Darwin's. This encouraged Darwin, for the sake of priority, to write out his views for publication. As a result, on November 22, 1859, *Origin of Species* was published, and it became one of the most influential works ever written. Darwin argues two major points in his book: organisms change over time, and this change is brought about through natural selection. The first point

was better accepted, at least initially, than the second. Interestingly, the word "evolution" does not appear in his book.

Huxley, Thomas H. *Man's Place in Nature and Other Anthropological Essays.* New York: Hurst, [1863] 1894; 274 pp.
Huxley (1825-1895) was a biologist and a vigorous champion of evolutionary theory. He was a member of Charles Darwin's inner circle, and for his outspoken efforts on Darwin's behalf, Huxley became known as "Darwin's bulldog." It was also Huxley who coined the word "agnostic" to describe someone who neither affirms nor denies the existence of God. One of Huxley's chief interests was comparative anatomy. While Darwin decided not to broach the issue of human evolution in *Origin of Species*, Huxley did so in a small book, *Man's Place in Nature*, first published in 1863. The book begins with a treatise on the natural history of manlike apes—chimpanzees, gibbons, gorillas, and orangutan—as well as a brief section on alleged African cannibalism. Huxley then discusses the anatomy of humans in relation to that of other animals, concluding "that Man is, in substance and in structure, one with the brutes." He then turns to the fossil remains of humans, mostly of the Neanderthal type, of which there were relatively few known in 1863. The 1894 edition, cited here, includes three additional essays on the issue of human history: two addressing the topics of ethnology and language, and a third pursing the roots of "Aryan" peoples.

White, Ellen G. *Spiritual Gifts: Important Facts of Faith in Connection with the History of Holy Men of Old.* Battle Creek, MI: Steam Press of the Seventh-day Adventist Publishing Association, 1864; 304 pp.
White (1827-1915) was co-founder and spiritual leader of the Seventh-day Adventist Church. Beginning with creation and the flood, *Spiritual Gifts* recounts Old Testament stories of faith. White describes the newly created earth as one with "hills and mountains, not high and ragged as they now are, but regular and beautiful in shape." Trees of the time, she writes, "were many times larger" than trees today, while Adam "was more than twice as tall as men now living upon the earth." But after Adam and Eve's fall into sin, the creation began to deteriorate. The worst sin of time, White asserts, was "the base crime of amalgamation of man and beast." More than any other, this sin "called for the destruction of the race by the flood." The flood waters,

which covered the whole earth, came from the skies above and from the "great deep" beneath. "Jets of water . . . burst up from the earth with indescribable force," she writes, "throwing massive rocks hundreds of feet into the air." After the flood the earth had lost its pristine beauty. She devotes an entire chapter to "Disguised Infidelity," by which she means the views of "Infidel geologists [who] claim that the world is very much older than the Bible record makes it." According to White, "creation week was only seven literal days, and . . . the world is now only about six thousand years old." Moreover, anyone who seeks to accommodate Genesis and geology through the day-age theory is deluded. *Spiritual Gifts*, along with later versions of this work under different titles, was considered by George McCready Price to be divinely inspired and provided the motivation for his flood geology. Price's ideas, in turn, were incorporated into Whitcomb and Morris' *The Genesis Flood*, from which creation science emerged. Thus, White was, in a sense, the mother of modern creationism.

Argyll, the Duke of [George Douglas Campbell]. *The Reign of Law.* First American edition from the fifth London edition. New York: Routledge & Sons, 1868; 462 pp.

The Duke of Argyll (1823-1900) was a government official who took an intense interest in the the controversy over Darwinian evolution. *The Reign of Law* is, in large part, a response to Darwin's theory of evolution by natural selection, of which the Duke is highly critical. *The Reign of Law* begins by examining what the author means by "the Supernatural," which he says is no different from "the Natural." Indeed, all events are outworkings of what Argyll calls the "Reign of Law," which is God's way of seeing his purposes fulfilled. Argyll is not uncomfortable with the notion of transmutation, so long as it is interpreted as the outworking of the Reign of Law. Darwin's natural selection is an impersonal, materialistic process, and therefore rejected by Argyll. Argyll sees various "contrivances," or adaptations, as arising due to the outworking of this law. He is particularly impressed in this regard by the flight of birds. He believes that the existence of beauty, as in hummingbird plumages, can be accounted for only on the basis of the outworking of the Reign of Law. The same reasoning is used to explain the existence of human mind and politics. Darwin and his allies found Argyll's arguments clever but unconvincing.

Mivart, St. George. *On the Genesis of Species.* London: Macmillan, 1871; 314 pp.
Mivart (1827-1900), a protégé of Thomas Huxley, convert to Catholicism, and lecturer in zoology at St. Mary's Hospital, Paddington, became a friend of Charles Darwin. At first, Mivart expressed acceptance of Darwin's theory of natural selection. As time went on, his support waned, until in 1871, just prior to the release of Darwin's *The Descent of Man*, Mivart published *On the Genesis of Species,* a withering critique of Darwin's views. Darwin was stung by his erstwhile friend's about-face and wanted nothing more to do with him. Mivart contends that natural selection could never accomplish what Darwin intended for it to do in the years allotted for the existence of the earth; he disputes Darwin's theory of pangenesis, he ridicules the notion that an organ like a wing could gradually evolve, and he eschews Darwinian metaphysics. According to Mivart, when species change, they do so suddenly, transforming into new, fully adapted forms due to the outworking of a mystical inner force; natural law shapes the world; and divine design and purpose are evidenced by the natural order.

Darwin, Charles. *The Descent of Man and Selection in Relation to Sex.* Two volumes. New York: Appleton, 1872; 845 pp.
Darwin had virtually ignored the question of human evolution in *Origin of Species,* but many of his supporters and critics had spoken out on human history, and Darwin felt compelled to issue his own pronouncements on the topic. According to the introduction to *The Descent of Man,* "The sole object of this work is to consider, firstly, whether man, like every other species, is descended from some preexisting form; secondly, the manner of his development; and thirdly, the value of the differences between the so-called races of man." Darwin begins by examining the evidence for human affinities with lower animals on the basis of homology. He then compares the mental powers of humans with lower animals, speculates as to the mode of development of humans from lower forms, discusses human intellectual development from primeval times, and examines the evidence for human genealogy. Darwin notes that "it has seemed to me highly probable that sexual selection has played an important part in differentiating the races of man"—three quarters of *The Descent of Man* is devoted to the evidence for sexual selection in a variety of organisms. Darwin concludes this work with the statement that "Man

still bears in his bodily frame the indelible stamp of his lowly origin."

Agassiz, Louis. *Methods of Study in Natural History.* Boston, MA: Osgood, 1878; 319 pp.
Agassiz (1807-1873), a Swiss immigrant to the United States, was one of the most influential scientists in nineteenth century America. A professor of natural history at Harvard University, Agassiz was committed to the view of multiple creations by God over long periods of time. Throughout his life, he remained opposed to the theory of evolution. He is perhaps best known for his studies on glaciation, a process he believed was responsible for much of the geologic evidence that others had attributed to the action of Noah's flood. Agassiz' *Methods of Study in Natural History* contains a series of essays designed to complement and place in popular form sentiments expressed in his earlier *Essay on Classification* (1859). In addition to providing students with information on how "scientific truth has been reached," he also takes the "opportunity to enter my earnest protest against the transmutation [evolutionary] theory, revived of late with so much ability, and so generally received." He says that naturalists enamored with this theory "are chasing a phantom." After a discussion of the history of natural science, he takes up the topic of classification, which he applies to the various groups of animals. He discusses the characteristics of taxonomic families, genera, and species, and concludes that "Species were created by God." Using the consistency of organisms in coral reefs over long periods of time, Agassiz argues for the immutability of species. As for homologies, Agassiz insists that they represent the creative expressions of "the Almighty Mind" on general themes and that they do not constitute evidences for evolutionary relationships. He closes with discussions of alternation of generations, ovarian eggs, and the relationship between embryology and classification.

Butler, Samuel. *Evolution Old and New: Or the Theories of Buffon, Dr. Erasmus Darwin and Lamarck, as Compared with that of Charles Darwin.* Third edition. London: Fifield, [1879] 1911; 340 pp.
The novelist Butler (1835-1902) became a Darwinian after reading *Origin of Species.* He knew Darwin personally, and his father and Darwin had been friends at Cambridge University. When Butler read Mivart's *On the Genesis of Species*, he became skeptical of natural

selection. Moreover, he felt that Darwin had appropriated his evolutionary outlook from Erasmus Darwin, Buffon, and Lamarck without giving these predecessors due credit. *Evolution Old and New* is Butler's attempt to right this perceived wrong. He agrees with Erasmus Darwin that mind is the real force for change in nature, not matter, as grandson Charles believed. Butler favors Lamarckian concepts as well. Butler and Darwin ended up as bitter enemies over Butler's book, as well as over an unsubstantiated claim by Butler that he had been deliberately slighted by Darwin.

Guyot, Arnold H. *Creation: Or the Biblical Cosmology in Light of Modern Science.* New York: Charles Scribner, 1884; 136 pp.
Guyot (1807-1884) was a professor of geology and physical geography at the College of New Jersey, and he was one of the only prominent scientists in the United States who had not accepted wholesale evolution by the 1880's. Born in Switzerland, Guyot introduced the discipline of scientific geography to America. In *Creation,* Guyot rejects liberal interpretations of the Bible and accepts the view that God was the originator of the universe. Guyot believes that Genesis speaks of three primary creations: of matter, of animals, and of humans—none of these entities could have materialized through evolution. Moreover, "Evolution from one of these orders into the other—from matter into life, from animal life into the spiritual life of man—is impossible." He grants the possibility, however, that evolution occurred *within* each of these orders as the result of natural law. He believes that the Genesis creation stories exhibit "wonderful caution in the statement of facts, which leaves room for all scientific discovery." Guyot posits that God communicated the story of creation to the writer of Genesis by means of pictorial visions, with each creation day representing one geologic age.

White, Andrew Dickson. *A History of the Warfare of Science with Theology in Christendom.* Two volumes. New York: Appleton, 1896; 989 pp.
White (1832-1918) was the founder and president of Cornell University. This two-volume work is dedicated to the memory of Ezra Cornell, chief benefactor of the university. White was a liberal Christian who disdained sectarianism. He also deplored what he saw as "a struggle between Science and Dogmatic Theology." He was convinced "that Science, though it has evidently conquered Dogmatic Theology

based on biblical texts and ancient modes of thought, will go hand in hand with Religion." His voluminous history addresses many disciplines, including evolutionary biology, geography, astronomy, geology, Egyptology, Assyriology, prehistoric archaeology, anthropology, ethnology, meteorology, chemistry, physics, medicine, public health, mental illness, language, economics, and biblical interpretation. Virtually every page is framed in the context of his warfare outlook, with science winning every battle over naive religion. While White's work has been very influential, science historians of today typically reject his simplistic warfare metaphor, positing instead a more positive dynamic between science and theology through time.

Scofield, Cyrus I. (ed.). *The Scofield Reference Bible.* New York: Oxford University Press, 1909; 1362 pp.

Scofield (1843-1921) was a lawyer, Congregational minister, and conservative biblical scholar. *The Scofield Reference Bible* consists of the King James version, accompanied by copious reference notes. Scofield's major interpretive scheme is dispensationalism, the idea that God establishes different covenants with his people at different times. In an annotation, Scofield suggests that the "heaven and the earth" of Genesis 1:1 "refers to the dateless past, and gives scope for all geologic ages." Later he posits: "Relegate fossils to the primitive creation, and no conflict of science with the Genesis cosmogony remains." Scofield's comments thus gave credence to the gap theory, which postulates two creations, the first of which is only briefly alluded to in Genesis 1:1; the second receives more detailed coverage in the remainder of the chapter. It is estimated that more than ten million copies of *The Scofield Reference Bible* were sold. It is little wonder, then, that the gap theory gained such popularity among fundamentalists of the early twentieth century. Indeed, popular evangelists like Harry Rimmer preached this view. A revised, 1967 edition of *The Scofield Reference Bible* omitted the earlier, more overt comments supporting the gap theory.

Wright, G. Frederick. *Origin and Antiquity of Man.* Oberlin, OH: Bibliotheca Sacra, 1912; 547 pp.

After graduating with a degree in theology from Oberlin College, Wright (1838-1921) became a Congregationalist pastor in Bakersfield, Vermont. During his pastorate, he grew to be well acquainted with the glacial deposits in the region, and eventually he became a respected geologist. He was particularly impressed with the views of Asa Gray, a Christian Darwinist at Harvard who later became his close friend. Wright and Gray accepted Darwinian natural selection but believed that God was involved in the production of the variation worked upon by selection. Eventually, Wright returned to Oberlin College, where he taught at the Theological Seminary. As he grew older, he became more critical of Darwinism. His *Origin and Antiquity of Man* argues that life is not older than fifty million years and that humans appeared less than fifteen thousand years ago. It also posits a possible biological link between humans and other primates but insists that the mental and moral capacities of humans were created by God. Changes in organisms occurred over time by "paroxysms of nature." Ironically, Wright's *Origin and Antiquity of Man* was published in the same year as his essay "The Passing of Evolution," in *The Fundamentals*—a series of tracts from which fundamentalism was born. While Wright's sentiments in *The Fundamentals* are phrased more conservatively than those in his book, they deviate dramatically from fundamentalist notions of late twentieth century creationism—notions that more closely resemble the biblical literalism of George McCready Price.

Price, George McCready. *The New Geology.* Mountain View, CA: Pacific Press, 1923; 726 pp.

Price (1870-1963) was a Seventh-day Adventist teacher and self-proclaimed geologist. He had virtually no formal training in geology, except for a mineralogy course at a teacher training school, but he was a capable writer, and he published hundreds of articles and dozens of books on creationism. He was, unquestionably, the most important creationist of the early twentieth century. *The New Geology* is written as a college text. Topics addressed include physical geology, stratigraphy, the fossil record, and theoretical geology. Two places in the book, Price enunciates his "great law of conformable stratigraphic sequence . . . which is by all odds the most important law ever formulated with reference to the order in which the strata occur: Any kind of fossiliferous beds whatever, 'young' or 'old,' may be found occurring conformably on any other fossiliferous beds, 'older' or 'younger.'" Thus, Price rejects the notion of a geologic column. He also dispenses with the concepts of thrust faulting and continental glaciation. He proposes, instead, that the flood described in Genesis was responsible for most of the earth's geologic formations. Future creationists, including John C. Whitcomb and Henry M. Morris, authors of *The Genesis Flood*, based their views on those of Price. Price's commitment to flood geology was rooted in the nineteenth century writings of Ellen G. White, prophetess to the Seventh-day Adventist Church.

Dobzhansky, Theodosius. *Genetics and the Origin of Species.* New York: Columbia University Press, 1937; 364 pp.
An emigrant from Russia, Dobzhansky (1900-1975) was a professor of genetics at the California Institute of Technology and a Christian. He had researched genetic variation in natural populations of fruit flies, greatly extending the database of population genetics. In *Genetics and the Origin of Species* he applies that database to the problem of evolutionary mechanisms. Specifically he asks how new species come about. He begins by addressing the problem of why are there so many types of organisms. He then examines gene and chromosome mutations as engines of variation, describes variation in natural populations, looks at the role of natural selection in culling variants, addresses cataclysmic speciation through autopolyploidy and allopolyploidy, highlights the significance of isolating mechanisms for the speciation process, evaluates the importance and role of hybrid sterility, and examines what is meant by the term "species." *Genetics and the Origin of Species* was one of the most influential books on evolutionary biology in the twentieth century. Along with books by Ernst Mayr (1942) and George Gaylord Simpson (1944), it played a key role in the emergence of neo-Darwinism.

Goldschmidt, Richard. *The Material Basis of Evolution.* New Haven, CT: Yale University Press, 1940; 436 pp.
Goldschmidt (1878-1958) was a professor of zoology at the University of California, Berkeley. His research interests included sex determination, physiological genetics, and evolutionary biology. In *The Material Basis of Evolution* Goldschmidt says he does not "agree with the viewpoint of the textbooks that the problem of evolution has been solved as far as the genetic basis is concerned." He is critical of "the strictly Darwinian view" and challenges its proponents to show how that view explains the gradual evolution of things like mammal hair, bird feathers, body segmentation of arthropods and vertebrates, teeth, mollusc shells, compound eyes, blood circulation, poison apparatus of snakes, and chemical differences between such

molecules as hemoglobin and hemocyanin. He asserts that subspecies are not incipient species but rather evolutionary dead ends. "The step from one species to another," he posits, "requires another evolutionary method than that of sheer accumulation of micromutations." In place of the Darwinian perspective, Goldschmidt argues that evolution occurs as the result of rare macromutations which instantaneously produce new features. He calls organisms that express these new features "hopeful monsters" and lists Manx cats, *Archaeopteryx*, flounders, and dachshunds as possible examples. Goldschmidt's hopeful monster theory was broadly rejected by the neo-Darwinists of his day, but some of his views are now receiving a respectful second look.

Huxley, Julian. *Evolution: The Modern Synthesis.* New York: Harper & Brothers, 1942; 645 pp.
Huxley (1887-1975) was a grandson of Thomas H. Huxley, a fellow of the Royal Society and a leading evolutionary biologist. It was Julian Huxley who coined the term "modern synthesis," which refers to the theoretical integration of genetics, systematics, paleontology, and evolutionary biology, and which led to neo-Darwinism. This book provides a comprehensive synthesis of systematics and genetics. It begins with a consideration of the theory of natural selection and adaptation, followed by a discussion of mutation, Mendelian genetics, and evolutionary genetics. The concept of the species is addressed, with particular attention paid to geographical features of speciation. Other aspects of speciation are considered as well, including local versus geographical differentiation, ecological divergence, genetic divergence, reticulate differentiation, and speciation in relation to taxonomy. Adaptation is discussed further in the context of natural selection, as are evolutionary trends and evolutionary progress. While Huxley is wary of naive conceptions of evolutionary progress, he nevertheless believes that "Progress is a major fact of past evolution," at least in a limited way. This book was important to the establishment of neo-Darwinism in England.

Mayr, Ernst. *Systematics and the Origin of Species.* New York: Columbia University Press, 1942; 334 pp.
Mayr (b. 1904), an ornithologist and animal systematist, was a curator at the American Museum of Natural History and a professor of zoology at Harvard University. His research in systematics led to the

development of the *biological species concept.* Along with works by Theodosius Dobzhansky and George Gaylord Simpson, Mayr's *Systematics and the Origin of Species* helped give birth to neo-Darwinism. Specifically, he synthesizes evidence from genetics and ecology with that of his own specialty, systematics, to build a coherent view of speciation in animals. He begins by describing the methods and principles of systematics, as well as the variation and meaning of taxonomic characters. He then examines geographic variation, biological species, polytypic species and their significance in nature and to systematics, the role of the species in evolution, the paucity of sympatric speciation, the biology of the speciation process, and macroevolution and the meaning of higher taxonomic categories. Mayr is generally credited with developing many of our present notions of how new species form. He has played a major role in twentieth century evolutionary biology.

Simpson, George Gaylord. *Tempo and Mode in Evolution.* New York: Columbia University Press, 1944; 237 pp.

Simpson (1902-1984), a professor at Columbia University and later at Harvard, and curator of paleontology at the American Museum of Natural History, was one of the twentieth century's preeminent paleontologists. His *Tempo and Mode in Evolution* represents an attempt to synthesize data from paleontology and genetics. This was a bold move on the part of Simpson, because traditionally paleontologists and geneticists operated in different conceptual worlds. Evolutionary theory was important to both groups but meant different things. Geneticists saw evolution as a change in the frequencies of genes in populations, while paleontologists saw evolution reflected in the fossil record as a progression of life-forms. Simpson realized that paleontologists, who typically focused on the *course* of evolution, needed to pay closer attention to genetics in order to understand the *mechanism* of evolution. In *Tempo and Mode*, Simpson begins by examining the rates of evolution implied by changes in fossils through time. He then examines various determinants of evolution, including variability, mutation rate, generation time, and natural selection. He defines "microevolution," "macroevolution," and "megaevolution," often poorly defined and misused terms. He evaluates the evidence for evolutionary trends, slow and fast-paced change, and adaptation. He finishes with a discussion of modes of change: speciation, phyletic evolution, and quantum evolution. *Tempo and Mode*, along with books by Theodosius Dobzhansky and Ernst Mayr, provided a major contribution to the evolutionary synthesis from which neo-Darwinism emerged.

Clark, Harold W. *The New Diluvialism.* Angwin, CA: Science Publications, 1946; 222 pp.
Clark (1891-1986), a Seventh-day Adventist and former student of George McCready Price, taught biology and geology at Pacific Union College. Initially, like Price, Clark believed that the fossil sequence was a fabrication, and he used Price's *The New Geology* as a textbook for his course in earth science. Clark discovered, however, that the petroleum industry used the fossil sequence as a practical tool to locate oil. Field work also convinced him that, contrary to the views of Price, thrust faulting and continental glaciation had occurred. After rethinking his views on geology, he wrote *The New Diluvialism*. While Clark remained completely committed to flood geology, he explained the orderliness of fossils, which he now accepted, in terms of his new "ecological zonation theory." In this view, the pre-flood world consisted of altitudinal life zones, similar to the ecological life zones in mountainous regions today. Thus, different organisms inhabited different levels. As the flood waters rose, progressively higher and higher life zones were buried. The fossil record, then, reflects this progressive burial process and provides a picture of the zonation of pre-flood life. *The New Diluvialism* also discusses biological change, the fossil history of humans, the age of the earth, and other topics under the framework of what Clark calls "neo-creationism," a perspective somewhat more open to scientific evidence than Price's view. Clark's ecological zonation theory continues to inspire the thinking of flood geologists today.

Velikovsky, Immanuel. *Worlds in Collision.* New York: Macmillan, 1950; 401 pp.
Velikovsky (1895-1979) was a Russian-born physician and psychoanalyst whose books stirred considerable controversy. In the preface to *Worlds in Collision*, Velikovsky warns readers that if Newton and Darwin "are sacrosanct, this book is a heresy." His purpose, he writes, is "to show that two series of cosmic catastrophes took place in historical times, thirty-four and twenty-six centuries ago, and thus only a short time ago not peace but war reigned in the solar system." He postulates that the planets have traveled in "their present orbits for

only a few thousand years," that Venus was once a comet originating through a violent eruption on Jupiter, and that comets originated through interactions between Venus and Mars. He claims that the orbit of the earth, the length of the year, the length of the day, and the terrestrial axis changed repeatedly. He also postulates that "electrical discharges took place between Venus, Mars, and the earth" that impacted the earth in many ways, resulting in mountain building, volcanic activity, earthquake activity, the creation of features usually attributed to continental glaciation, the Genesis flood, the appearance of manna as food for the wandering Israelites, the parting of the Red Sea, Joshua's long day, and other physical phenomena and biblical miracles. Velikovsky appeals to the memory of these events alluded to in epic tales and folklore. Most creationists and evolutionists rejected his interpretations as fantastic, though creationists admired his affirmation of the biblical stories and the fact that he embraced catastrophism.

Ramm, Bernard. *The Christian View of Science and Scripture.* Grand Rapids, MI: Eerdmans, 1954; 256 pp.

Ramm (1916-1993) was a professor of systematic theology and Christian apologetics at American Baptist Seminary of the West. He disagreed with the narrow views taken by conservative Christians on issues of science and faith during the first half of the twentieth century. He believed that evangelical Christians had turned their backs on "the tradition of the great and learned evangelical Christians who have been patient, genuine, and kind and who have taken great care to learn the facts of science and Scripture." He was particularly unhappy with the views of George McCready Price and others of Price's interpretive bent. Ramm's book begins with the suggestion that a harmony of Christianity and science is imperative. This is followed by an analysis of the conflict between science and theology, and some specific problems inherent to this conflict. Various topics in astronomy, geology, biology, and anthropology are systematically addressed in view of the biblical record. Ramm favors an interpretation of Genesis 1 he calls the "pictorial day theory," which suggests that the "main purpose of Genesis is theological and religious." In this view, the six days of creation "are *pictorial-revelatory* days, not literal days nor age-days." In other words, days were used metaphorically by God to communicate the truth about creation to the author of Genesis 1. Ramm rejects the notion of a universal flood, accepts the earth's great antiquity, and believes in progressive creationism. *The Genesis Flood* by Whitcomb and Morris, in part, was a response to Ramm, whom they felt had sold out to modernism.

Popper, Karl R. *The Logic of Scientific Discovery.* New York: Basic, 1959; 480 pp.
Popper (1902-1994) was one of the twentieth century's most influential philosophers of science. *The Logic of Scientific Discovery* was first published in 1934 under the German title *Logik der Forschung*. Popper first introduces the logic of science, discussing inductivism, deductivism, experience, empiricism, objectivity, subjectivity, and scientific methodology. He then examines some structural components of "a theory of experience," followed by his views on falsifiability, empiricism, testability, simplicity, and probability. In one chapter he makes some observations on quantum theory and its implications for the interpretation of scientific data. Finally, he examines the process of scientific verification through hypothesis testing. In the context of the creation/evolution controversy, Popper's view that falsifiability is a criterion for determining whether a hypothesis has scientific merit is especially noteworthy. Hypotheses in paleontology, historical geology, evolutionary biology, and other historical disciplines are said by creationists to exist outside the realm of science, because they deal with the past and, therefore, are beyond the realm of falsification. While Popper's view is considered important, many scientists do not consider falsifiability to be an absolute criterion of what constitutes science.

Teilhard de Chardin, Pierre. *The Phenomenon of Man.* New York: Harper & Brothers, 1959; 318 pp.
Teilhard de Chardin (1881-1955), a French Jesuit priest and paleontologist, moved to the United States from leadership positions in geology in France. In the United States he continued his research at the Wenner-Gren Foundation. He played an important role in the discoveries of Peking Man and the infamous Piltdown Man. He had a mystical view of evolution, the method by which he believed God created. He was, thus, a theistic evolutionist. *The Phenomenon of Man* is introduced by Sir Julian Huxley, who reviews Teilhard's life and extols his work. Teilhard notes in his preface that the book should be read as a scientific, not a theological or metaphysical, work. After examining the nature of matter and the early earth, he looks at life, its evolu-

tionary beginnings, characteristics, and diversity. This is followed by a discussion of the development of consciousness, especially in relation to human evolution. A concluding section assesses some of the more philosophical aspects of human existence, including personhood, love, and religion. An appendix remarks on the existence of evil in an evolutionary world. According to Teilhard, "man is seen not as a static centre of the world—as he for [so] long believed himself to be—but as the axis and leading shoot of evolution, which is something much finer."

Whitcomb, John C., Jr., and Henry M. Morris. *The Genesis Flood: The Biblical Record and Its Scientific Implications.* Philadelphia, PA: Presbyterian and Reformed, 1961; 518 pp.
When they wrote this book, Whitcomb (b. 1924) was a professor of Old Testament at Grace Theological Seminary, while Morris (b. 1918) was head of the Department of Civil Engineering at Virginia Polytechnic Institute. Both authors were committed to the position "that the Bible is the infallible Word of God, verbally inspired in the original autographs." Their interpretations of both the Bible and geology in *The Genesis Flood* are consistent with this affirmation. The book begins by arguing that the flood covered the entire earth. Arguments against this proposition are systematically addressed and discounted. Multiple catastrophism, uniformitarianism, and local flood theories are also examined and dismissed. Various geological phenomena are discussed in view of flood geology, and a scriptural framework for historical geology is developed. A final chapter addresses problems with young-earth flood geology, including evidence from radiometric dating, tree rings, coral reefs, deep-sea sediments, evaporite deposits, varves, and petroleum deposits. None of these problems is considered insurmountable. One appendix addresses the issue of paleontology and the Edenic curse, while a second considers when the flood occurred in relation to the Genesis genealogies. Whitcomb and Morris' geological interpretations follow those of George McCready Price in many respects. Publication of *The Genesis Flood* in 1961 marked the beginning of the modern creation science movement in America.

Kuhn, Thomas S. *The Structure of Scientific Revolutions.* Second edition. Chicago, IL: University of Chicago Press, [1962] 1970; 210 pp.

Kuhn (b. 1922), a professor at the Massachusetts Institute of Technology, studied theoretical physics but shifted his focus to the history and philosophy of science. *The Structure of Scientific Revolutions* is often considered one of the landmark contributions to scientific thought during the twentieth century. Kuhn begins by asserting the importance of the history of science in the quest to understand how science operates. History demonstrates, he says, that science makes significant progress, not by gradually accumulating ever-increasing piles of evidence, but through revolutions. Most science, even good science, is done in the context of what he calls *normal science*. Normal science is based in a particular system of thought, or *paradigm*. Sooner or later, anomalous evidence, incompatible with the dominant paradigm, accumulates to the point that a crisis develops. The crisis forces a reexamination of old evidence and the collection of new evidence, sometimes with the advantage of new tools. Finally, a *paradigm shift* occurs, an event of significant scientific progress. Thus, he says, science proceeds like biological evolution: "The net result of a sequence of such revolutionary selections, separated by periods of normal research, is the wonderfully adapted set of instruments we call modern scientific knowledge." Textbooks, he believes, are in large part responsible for the false view that science progresses gradually, like "the addition of bricks to a building." Creationists sometimes use Kuhn's perspective to predict that science will eventually be forced by evidence to take seriously a theistic view of origins.

Sagan, Carl. *Cosmos*. New York: Random House, 1980; 365 pp.
Sagan (1934-1997) was a professor of astronomy and space sciences at Cornell University. He played leading roles in the Mariner, Viking, and Voyager space expeditions, and he won many awards, including a Pulitzer Prize for his writing. His immensely popular public television series, *Cosmos*, also aired in 1980. Sagan's book is not so important for the new insights it provided—it was, after all, a popular work—but along with the companion television series, this book had a profound influence on the public and its perceptions of the universe. Theists were distressed, however, by Sagan's unabashed scientism. He begins both the television series and Chapter 1 of the book with the sentence: "The Cosmos is all that is or ever was or ever will be." Throughout the book religion is treated as an outdated myth now thoroughly discredited and superseded by science. Thus, for some observers, *Cosmos* came to stand for science at its worst: imperialistic,

positivistic, and materialistic. Whatever they thought about his philosophy, nearly everyone was impressed by Sagan's communicative style. The blend of historical insights, contemporary interpretations of the universe, and superb graphics makes *Cosmos* a visual and intellectual feast.

Chapter 3

HISTORY OF THE CREATION/EVOLUTION CONTROVERSY

Studies of how people have conceptualized creation or evolution—histories of histories—have appeared with increasing frequency in recent years. These works show that religious belief, social context, and political reality shape the ways in which people interpret evidence from the past. They also raise the interesting question of how the ideas of today will look to future generations.

Azar, Larry. *Twentieth Century in Crisis: Foundations of Totalitarianism.* Dubuque, IA: Kendall/Hunt, 1990; 317 pp.
Azar is a conservative philosopher. In this book he argues that evolutionism was the force behind the twentieth century totalitarianism of Adolf Hitler. Azar writes that Hitler's racism was rooted in his evolutionary views, and that he murdered millions of people because he believed he was simply carrying out the imperatives of natural selection. Azar argues that Darwin's views became popular because the way to acceptance had been paved philosophically. Indeed, Azar points out, it was philosophers who first made the word "evolution" popular. Also of interest to the creation/evolution debate, Azar discusses the argument from design and provides several examples of apparent design in organisms.

Bergman, Jerry. *The Criterion: Religious Discrimination in America.* Richfield, MN: Onesimus, 1984; 80 pp.
Bergman, a professor of educational foundations and inquiry at Bowling Green State University, was denied tenure because of his creationist beliefs. This book reviews the cases of more than one hundred individuals who, because of their creationist views, experienced discrimination in the workplace and in academia. Most of

the interviewees had Ph.D. degrees, and all had at least master's degrees. An appendix features an article by Luther D. Sunderland detailing Bergman's own case.

Biagioli, Mario. *Galileo, Courtier: The Practice of Science in the Culture of Absolutism.* Chicago, IL: University of Chicago Press, 1993; 402 pp.
Biagioli, an associate professor of history at the University of California, Los Angeles, places the career of Galileo in the context of his role as a royal courtier. According to Biagioli, the seventeenth century court played a significant role in fostering science. Moreover, Galileo's 1633 fall from grace in the papal court of Urban VIII was as much the result of a courtly political struggle as it was over the conflict between his Copernican cosmology and Thomistic theology. This book covers only two and one-half decades of Galileo's career and is, therefore, not intended as a biography. It does, however, highlight the significant role played by politics in matters of science and faith, a role that continues today.

Blackmore, Vernon, and Andrew Page. *Evolution: The Great Debate.* Oxford: Lion, 1989; 192 pp.
The intent of the authors of this book is to tell the history of the theory of evolution and to place this history in the context of political and social forces, religious conflict, and an intense human desire to understand origins. They trace evolutionary views from the early Greek philosophers to the present and finish their story with a discussion of the views of prominent nineteenth and twentieth century personalities such as Thomas Huxley, Julian Huxley, Bertrand Russell, Richard Leakey, and Jacques Monod. Although the authors reject naturalistic evolutionism as a worldview, they also reject creation science, preferring to simply believe that God is the originator and sustainer of natural phenomena and that humans will never be able to discover God using the methods of science. This well-illustrated book is written in a popular, engaging style.

Bowler, Peter J. *The Eclipse of Darwinism: Anti-Darwinian Evolution Theories in the Decades Around 1900.* Baltimore, MD: The Johns Hopkins University Press, 1983; 291 pp.
Bowler is a historian at The Queen's University, Belfast. In this well-referenced work, he recounts the controversy that surrounded Darwin's

theory of evolution in the decades following publication of *Origin of Species.* Bowler details non-Darwinian theories of evolution including Lamarckism, mutationalism, and orthogenesis. He points out that Darwin's theory of evolution by natural selection became the generally accepted view only after development of the "modern synthesis" achieved in the 1930's by geneticists such as Sewell Wright and Theodosius Dobzhansky, and by paleontologists and field biologists such as George Gaylord Simpson and Ernst Mayr.

_____. *The Non-Darwinian Revolution: Reinterpreting a Historical Myth.* Baltimore, MD: The Johns Hopkins University Press, 1992; 238 pp.
Bowler argues that although publication of Darwin's theory accelerated acceptance of the concept of evolution in general, most people rejected Darwin's specific notion of change through natural selection. The concept of natural selection was offensive to many of Darwin's contemporaries who viewed the process as inhumane. Lamarckism, recapitulation theory, and goal-directed progressivism were much more popular. It was not until the 1930's, when principles of genetics were melded with those of Darwinian theory to form the so-called "new synthesis," that Darwin's theory of evolution by natural selection became generally accepted. Bowler also provides chapters on human evolution, social Darwinism, cultural evolution, and new historical perspectives on how evolutionary views have changed.

_____. *Darwinism.* New York: Twayne, 1993; 120 pp.
This book reviews the initially cool reception of Darwin's theory of evolution by natural selection and also comments on the role that Darwinism plays today in science. Bowler notes that people as diverse as botanist Asa Gray, astronomer William Hershel, and playwright George Bernard Shaw expressed serious reservations about aspects of Darwin's theory. Thomas Huxley, writes Bowler, concluded that humans are trapped in an amoral universe, and that this condition would eventually lead to a breakdown of civilization. Bowler suggests, however, that natural selection is not questioned in the same way today, largely due to the influence of government funding programs supportive of Darwinian explanations. Also discussed are Darwin's views on recapitulation, vestigial organs, and the human races.

Brooke, John Hedley. *Science and Religion: Some Historical Perspectives.* Cambridge, England: Cambridge University Press, 1991; 422 pp.

In this book Brooke offers numerous critical perspectives on historical encounters between science and religion. He rejects apologetic interpretations, highlighting instead the complexities, ambiguities, and ironies of these encounters. For example, he discusses how Georges Cuvier's careful description of the fossil record served to promote evolutionary theory, despite the fact that Cuvier himself was an ardent antievolutionist; how the geological theories of another antievolutionist, Charles Lyell, were used to buttress Charles Darwin's theories; and how divine design guru William Paley's descriptions of the perfect adaptations of organisms were appropriated as evidence for blind natural selection. In short, Brooke rejects simplistic characterizations, opting to focus on important subtleties that served as shaping forces in the science and religion controversy. Some of these subtleties were created by people less well known to twentieth century readers—people like William Whewell, Charles Hodge, John Henry Newman, and William Irons whom Brooke weaves into his perceptive narrative.

Brooks, John Langdon. *Just Before the Origin: Alfred Russel Wallace's Theory of Evolution.* New York: Columbia University Press, 1984; 284 pp.

Brooks, a program director at the National Science Foundation, is a former professor of biology at Yale University. In this book Brooks seeks to rescue the work of Alfred Russel Wallace from "undeserved obscurity." Brooks recounts Wallace's study of plants in England and Wales, his trip to Brazil in 1848, and his eight-year collecting trip to the Malay Archipelago starting in 1854. Like Charles Darwin, Wallace was influenced by the writings of Thomas Malthus and Charles Lyell. He also read Darwin's account of the voyage of the *Beagle.* Brooks details the evidence and events that led to Wallace's own theory of evolutionary change, first written in 1855; highlights differences between the theories of Wallace and Darwin; and points out that Darwin's views were influenced by those of Wallace, even though Wallace did not receive due credit. Brooks believes that Wallace's 1855 paper was what prodded Darwin to publish his own views in *Origin of Species* in 1859.

Cantor, Geoffrey. *Michael Faraday: Sandemanian and Scientist.* New York: St. Martin's, 1991; 359 pp.
Michael Faraday, the discoverer of electromagnetic induction and one of the great scientists of the nineteenth century, was a member of a small Christian sect, the Sandemanians. Sandemanians sought to restore Christianity to its pristine, first century form. According to science historian Cantor, we cannot understand Faraday, the scientist, without understanding Faraday, the Sandemanian. This book carefully examines Sandemanianism and its significant influence on Faraday's life and thought. In keeping with his faith, Faraday viewed nature as the Creation. He believed that science was simply his humble attempt to understand God's creation.

Cohn, Norman. *Noah's Flood: The Genesis Story in Western Thought.* New Haven, CT: Yale University Press, 1996, 240 pp.
Cohn is a professor emeritus of the University of Sussex and a fellow of the British Academy. In this carefully researched, well-illustrated book, Cohn traces the origin and development of the flood story and examines how it has been interpreted. He posits that the story was rooted in Mesopotamian mythology but that its original meaning was dramatically altered by the Genesis writers. The flood story gave hope to the Hebrews and provided assurance of salvation to early Christians. During the seventeenth, eighteenth, and early nineteenth centuries, the flood account played an important role in the developing science of geology. Eventually, Christians used the flood story as an interpretive device for explaining the geologic column in the context of biblical history. Fundamentalists today continue this tradition and even sponsor expeditions to look for a now grounded Noah's ark. Cohn concludes with an examination of the variety of interpretations placed on the flood story by secular scholars.

Corsi, Pietro. *Science and Religion: Baden Powell and the Anglican Debate.* Cambridge, England: Cambridge University Press, 1988; 346 pp.
The Reverend Baden Powell (1796-1860), a distinguished professor of geometry at Oxford, had wide-ranging interests in everything from optics to ecclesiastical history. Of great interest to him was the integration of theology, philosophy, and science. In his younger years, Powell believed that there was agreement between the biblical record and the geological record. In later years, however, Powell took a more

liberal theological stance and became more sympathetic to evolutionary views. This transition was due largely to advances in geological knowledge, as well as shifts in his political outlook and his higher critical studies of the Bible. Eventually he came to regard Genesis 1 as a mythical story designed to convey a religious message. Corsi's book provides a valuable study of a nineteenth century attempt to integrate science and faith.

Croft, L. R. *The Life and Death of Charles Darwin.* Lancanshire, England: Elmwood, 1989; 138 pp.
Croft begins by providing a brief overview of Charles Darwin's life. He then discusses the development of Darwin's theory of evolution and the events surrounding the publication of *Origin of Species* in 1859. Croft believes that, while Darwin eagerly sought the limelight for development of the theory of evolution by natural selection, several other individuals should be considered rightful contenders for this honor. One, of course, was Alfred Russel Wallace, whose views were made known publicly in joint presentation with those of Darwin. But there was also Pattrick Matthew, a horticulturalist, who included the essentials of natural selection in a book published in 1831. Finally, there was Edward Blyth, who wrote articles in the 1830's presaging Darwin's views. Croft reviews the variety of explanations for Darwin's illness, concluding that he was probably schizophrenic. He also contends that Darwin reestablished his link with Christianity just before he died. Most of the evidence for this contention is based in a story to this effect by a "Lady Hope" published in the early twentieth century.

Davis, Edward B. (ed.). *The Antievolution Pamphlets of Harry Rimmer.* Volume 6 of *Creationism in Twentieth-Century America,* Ronald L. Numbers, series editor. New York: Garland, 1995; 482 pp.
Harry Rimmer (1890-1952), a Presbyterian pastor and evangelist, was one of the most visible antievolutionists between the two world wars. Before becoming a Christian and taking ministerial training, Rimmer studied medicine. A gifted public speaker, Rimmer spoke widely at Christian churches, schools, Bible conferences, and youth camps. His Research Science Bureau was established to promote Bible-based research in biology, paleontology, archaeology, and anthropology, though its main function was to fund Rimmer's evangelistic activities. Rimmer defended the "gap theory" of Genesis 1.

This volume contains a biographical sketch of Rimmer and sixteen of his published pamphlets.

Degler, Carl N. *In Search of Human Nature: The Decline and Revival of Darwinism in American Social Thought.* New York: Oxford University Press, 1991; 400 pp.
Degler is an emeritus professor of American history at Stanford University and the winner of a Pulitzer Prize and a Bancroft Prize for his previous books. This well-referenced volume won him Phi Beta Kappa's Ralph Waldo Emerson Prize in 1991. Degler first examines the impact of Darwinism on human self-consciousness in relation to the earlier view that humans were "God's creatures." He also explores resultant nineteenth century biological justifications for racism and sexism, and the development of eugenics and intelligence testing. Clearly, he writes, the "influence of Darwinian ideas extended beyond the thought and attitudes of scholars; it shaped social policy as well." But by the 1920's, social scientists began to question the impact of biology on human behavior and intelligence. This shift was in large part due to the ideological and philosophical "belief that the world could be a freer and more just place" if biological theories of human behavior were refuted. By the 1950's and 1960's, however, advances in the understanding of the genetics of human behavior led social scientists to reembrace evolutionary explanations, minus the racist, sexist, and eugenic notions that characterized earlier accommodations to Darwinian biology. Degler concludes with an examination of the contemporary nature/nurture debate, particularly in relation to human consciousness, communication, cooperation, culture, and morals.

Desmond, Adrian. *The Politics of Evolution: Morphology, Medicine, and Reform in Radical London.* Chicago, IL: University of Chicago Press, 1989; 503 pp.
Charles Darwin delayed publication of his theory of evolution by natural selection for many years. This delay has been the subject of much speculation. Desmond's history of evolutionary theory in England during the 1830's provides an interesting perspective on this delay. Specifically, Desmond details the spread of Lamarck's "transformist" views from Paris to London, where these views took root among radical young medical doctors—"low-lifes" as Desmond calls them—who took pleasure in ruffling the feathers of the Tory medical establishment. Darwin, politically conservative and concerned

about his and his family's reputation, was well aware of how the young radicals had used earlier evolutionary ideas. According to Desmond, Darwin did not want to be identified with these subversive elements, so he held off publishing his own views. Desmond is a superb sleuth of history and a masterful writer. This outstanding book won him the History of Science Society's Pfizer Award in 1991.

_____. *Huxley: The Devil's Disciple.* London: Michael Joseph, 1994; 475 pp.
Desmond here portrays of the life of "Darwin's bulldog," Thomas Henry Huxley. This book is the first of two planned volumes on Huxley by Desmond, with this one detailing Huxley's life from the time of his birth in 1825, to his election as president of the British Association for the Advancement of Science in 1870. The second volume will cover Huxley's life from 1870 to the time of his death in 1895. Desmond recounts Huxley's apprenticeship with a Coventry doctor at age thirteen, his study of medicine at Charing Cross Hospital, his voyage to Australia on H.M.S. *Rattlesnake* as assistant surgeon and naturalist, his research on marine invertebrates, his early financial difficulties, and his eventual lectureship at the Government School of Mines. Desmond points out that while Huxley defended Darwin, this defense was based in Huxley's commitment to Darwin's nonteleological naturalism, not to his evolutionism. "In fact," he writes, "Huxley did not think in terms of *origins* at all. Geometry, not genealogy, fascinated him: the surreal beauty of nature's secret architecture." It was Ernst Haeckel's views that finally attracted Huxley to evolutionary theory, though he never fully subscribed to Darwin's theory of natural selection.

_____, and James Moore. *Darwin: The Life of a Tormented Evolutionist.* New York: Warner Books, 1991; 808 pp.
Scores of biographies on Charles Darwin have appeared, but this is probably the best. The book reads like a novel, yet it is thoroughly documented. The authors set out to "pose the awkward questions," questions such as: "How did such a wealthy Whig gentleman break the impasse and make evolution acceptable? How did he present it as underpinning middle-class values? Did he ever resolve the anti-thesis? . . .What of Darwin's own latter-day prejudices? . . . And how did grave Victorians observe [him]?" The answers provided will entertain, inform, and whet one's appetite for a deeper understanding of Charles

Darwin, the person, and the social milieu within which he developed his ideas.

Ellegard, Alvar. *Darwin and the General Reader: The Reception of Darwin's Theory of Evolution in the British Periodical Press, 1859-1872.* Chicago, IL: University of Chicago Press, 1990; 394 pp.
This book examines the impact of Darwin's theory of evolution on the British public during the first few years after publication of *Origin of Species.* Ellegard researched 115 British periodicals to collect this information. He evaluates the reception of Darwinism in the context of four major influences: politics, religion, science, and the philosophy of science. He explains that Darwin's major contribution was not the notion that organisms change over time, but the provision of a mechanism for that change, natural selection. Ellegard contends that although Darwin's theory did not oppose theology, it was seen by many people as a threat to their religious faith. He believes that the public's response to Darwin's theory was predicated primarily on the basis of its religious and ideological implications, not on the basis of scientific "fact."

Eve, Raymond A., and Francis B. Harrold. *The Creationist Movement in Modern America.* Boston, MA: Twayne, 1991; 234 pp.
Most books that examine creationism from the outside focus on its religious, philosophical, and scientific claims, or evaluate historical and political dimensions of the movement. Although these dimensions are not ignored in this book, sociologist Eve and anthropologist Harrold focus on the social and psychological dimensions of creationism. They suggest that creationism is less an argument over scientific interpretation than it is a competing worldview seeking to control the "means of cultural reproduction." Thus, creationism is not likely to be swept away by science, and the creation/evolution debate will only intensify in the future.

Geisler, Norman, in collaboration with A. F. Brooke, II, and Mark J. Keough. *The Creator in the Courtroom "Scopes II": The Controversial Arkansas Creation-Evolution Trial.* Milford, MI: Mott Media, 1982; 242 pp.
Geisler, a professor at Dallas Theological Seminary, served as lead defense witness at the 1981 Arkansas creationism trial. His book contains a detailed account of the trial, including a summary of the

testimony of each defense and prosecution witness, sample newspaper reports and magazine articles related to the trial, and key legal documents such as the text of Arkansas Act 590, Judge Overton's ruling, and the text of the related Louisiana Creation-Evolution Act. Geisler is a conservative creationist and his bitterness over the outcome of the trial is clearly evident; however, his book contains much useful information.

Gilkey, Langdon. *Creationism on Trial.* San Francisco: Harper & Row, 1985; 301 pp.
When he wrote this, Gilkey was a professor of theology at the University of Chicago Divinity School and had recently served as a theological witness for the American Civil Liberties Union at the 1981 creationist trial in Little Rock, Arkansas. The first two-thirds of the book consists of a delightfully written, personal account of pretrial proceedings and the trial itself, including a perceptive summary of the religious, historical, theological, philosophical, and scientific issues involved. In the last third, Gilkey analyzes the arguments used at the trial and reflects on the implications of creationism for modern society and contemporary religion. Appendices include important documents of the trial.

Glen, William. *The Road to Jaramillo: Critical Years of the Revolution in Earth Science.* Stanford, CA: Stanford University Press, 1982; 459 pp.
The theory of plate tectonics, which has become generally accepted during the last half of the twentieth century, has revolutionized the way geologists view the earth and its history. Plate tectonics has impacted many other scientific disciplines as well, including biogeography, paleontology, and planetary science. This book tells how this revolutionary theory developed and who developed it. It is based on interviews with over one hundred scientists involved in establishing the new paradigm. Part I examines the invention of potassium-argon dating during the early 1950's and its contribution to plate tectonic theory; Part II reviews the history of work in paleomagnetism, a crucial component of the plate tectonic puzzle; and Part III discusses the development of the sea floor spreading model, also crucial to the development of the new theory. This book provides an excellent case study of a major paradigm shift in scientific thought.

Glick, Thomas F. (ed.). *The Comparative Reception of Darwinism.* Chicago, IL: University of Chicago Press, 1988; 505 pp.
This book contains the proceedings of a conference by the same title held in 1972. The authors examine how Darwin's theory was received in nine countries and regions. Specifically, for each location, they deal with the sequence of events involved in promulgating Darwin's views; the sequence of arguments raised for and against his theory; societal factors that influenced the reception of his views; the social and political alliances of the friends and foes of Darwinism; the influence of educational level, profession, city, and specific region on acceptance of Darwinism; the influence of Darwinism on scientific research; and the impact of Darwinism on other areas of human endeavor. This is a scholarly work that will be of primary interest to historians of science.

Goldstein, Thomas. *The Dawn of Modern Science.* Boston: Houghton Mifflin, 1988; 297 pp.
This book is primarily a history of science before the time of Copernicus, told from a broad cultural perspective. Goldstein traces the roots of science back to the philosophers of ancient Greece. He shows how western science faded with the fall of the Roman Empire, then revived once again during the Renaissance. He examines how the earth was viewed by the inhabitants of Florence during the Renaissance, especially by Paolo Toscanelli, who conveyed his global concept of the world to Christopher Columbus. Goldstein examines the development of science at Chartres, where scholars saw nature as part of an overall divine plan. One such scholar, Thierry, taught his idea of God's "continuous creation," an ongoing beautification of the world. His successor, Conches, believed that it was not the Bible's role to teach the nature of things, a role reserved for science. Goldstein also addresses the contributions of Arabs, medieval mystics, alchemists, cathedral builders, physicians, and artists to the development of science during this period.

Gregory, Frederick. *Nature Lost? Natural Science and the German Theological Traditions of the Nineteenth Century.* Cambridge, MA: Harvard University Press, 1992; 341 pp.
Gregory, who has degrees in mathematics, theology, and the history of science, examines issues of science and faith in nineteenth century German-speaking Europe. After outlining his goals and methodology,

Gregory surveys the history of German protestant theology. He then reviews the works of several prominent theologians concerned about issues of science and religion: Friedrich Strauss, who developed a naturalistic, Darwinian view of the universe; Otto Zockler, an outspoken critic of Darwin who believed that the biblical story of creation was historical, and who attempted to find concordance between Genesis and geology; Rudolf Schmid, who believed in the doctrine of salvation but also refused to impose theological constraints on science; and Albrecht Ritschl, who completely separated religious and scientific thought. Gregory's outstanding study illuminates a topic that has remained largely unexplored by English-speaking scholars.

Hawkin, David, and Eileen Hawkin. *The Word of Science: The Religious and Social Thought of C. A. Coulson.* London: Epworth, 1989; 127 pp.
Charles Alfred Coulson (1910-1974) was a physical scientist interested in atomic and molecular structure. He taught at several institutions, including King's College and Oxford. Coulson was a Christian who was deeply interested in the relationship betweeen science and faith. His book *Science and Christian Belief* (1955) became well known to thoughtful Christians, and his views on the argument from design, "God of the gaps" thinking, and other theistic concerns were very influential. The Hawkins' book provides an important retrospective on Coulson's views on science and religion.

Hedtke, Randall. *The Secret of the Sixth Edition.* New York: Vantage, 1983; 136 pp.
Hedtke is a public school science teacher and a frequent contributor to the creationist literature. He believes that Darwin's health problems were caused by criticisms he received over his theory of natural selection promoted in the first edition of *Origin of Species.* In the sixth edition, Hedtke writes, Darwin disavowed his theory of natural selection, favoring instead a more Lamarckian view of change. According to Hedtke, Darwin's health improved after publication of the sixth edition because he had finally given in to his critics and accepted the falsity of evolution by natural selection. Darwin was not a scientist, opines Hedtke, nor did he understand science. Hedtke predicts that as the fossil record becomes better known, organisms once thought from a Darwinian perspective to have appeared only later in the history of the earth will show up at lower levels of the geologic column.

Herbert, David. *Charles Darwin's Religious Views: From Creationist to Evolutionist.* London, Ontario: Hersil, 1990; 104 pp.
Herbert is a secondary school history teacher and doctoral candidate at the University of Toronto. This book provides a history of Charles Darwin's views on religion. Herbert points out that Darwin grew up in a family of free thinkers, Unitarians, agnostics, and atheists. Moreover, his grandfather, Erasmus Darwin, who wrote the evolutionary book *Zoonomia,* and university teacher Robert Grant influenced Charles to think along evolutionary lines. But he was also close to several deeply religious people, including botanist John Henslow, geologists Adam Sedgwick and Charles Lyell, and his beloved wife Emma. Herbert shows that Darwin sought continually to balance the theistic and nontheistic influences on his life. While Darwin was an ardent proponent of naturalistic explanations, he never completely abandoned the possibility that there was some type of "First Cause" behind the first appearance of life. "Agnostic" seems to best describe the religious sentiments of Darwin in his later years.

Kalthoff, Mark A. (ed.). *Creation and Evolution in the Early American Scientific Affiliation.* Volume 10 of *Creationism in Twentieth-Century America,* Ronald L. Numbers, series editor. New York: Garland, 1995; 468 pp.
The American Scientific Affiliation (ASA) was organized in 1941 and continues to facilitate the discussion of science/faith issues among American evangelical scientists today. Kalthoff notes in his introduction that the "story of the ASA is largely the story of twentieth century evangelical scientists confronting questions of origins head-on, facing casualties, and living to tell about it." Early ASA members were conservative antievolutionists, but as time went on members became more open to standard scientific interpretations of biological and geological change. This volume provides a collection of early ASA documents that chronicle this shift in perspectives.

Larson, Edward J. *Trial and Error: The American Controversy over Creation and Evolution.* New York: Oxford University Press, 1985, 213 pp.
Larson has both a law degree and a Ph.D. in the history of science. This book recounts the history of the American creation/evolution controversy, with particular attention devoted to the legal battles. Larson first examines nineteenth century responses to Darwinism, in-

cluding those by American biologists, and looks at the impact of Darwinism on public opinion of the time. He notes that by 1900, most major textbooks included discussions of evolutionary theory, though not always to the exclusion of theistic viewpoints. After World War I high school enrollment was dramatically higher than before the war, with the consequence that many more young Americans were exposed to evolutionary theory. This increased exposure was of concern to fundamentalists, who responded by mounting an aggressive antievolution crusade. Larsen traces the history of this crusade through the legislatures, courts, schools, and textbooks, including the famous Scopes Trial in 1925 and the more recent Arkansas and Louisiana court battles. He assesses the impacts of the various interest groups that led to the decisions in each case and posits that the continued perseverance of these groups guarantees a long-term future for the creation/evolution controversy.

Lightman, Bernard. *The Origins of Agnosticism: Victorian Unbelief and the Limits of Knowledge.* Baltimore, MD: Johns Hopkins University Press, 1987; 249 pp.

Lightman traces the history of "agnosticism," the view that it is impossible to know if God exists, or the view that denies that we can know anything about God if such a being does exist. Agnosticism was developed by nineteenth century evolutionary naturalists such as Thomas Henry Huxley, Herbert Spencer, and John Tyndall. Lightman contends that although the founders of agnosticism were clearly antagonistic toward traditional religious beliefs and institutions, they were deeply religious people seeking to develop a new religious paradigm. In fact, agnostics were natural theologians who believed that nature itself was holy. At the heart of their faith was a confidence in "the ability of science to uncover the order of nature through an empirical study of the physical world." This was because they believed in the objective reality of the natural world and in the operation of causal determinism and lawful uniformity. This books sheds light on the origins of a prevailing philosophical view among natural scientists of the post-Darwin era.

Lindberg, David C. *The Beginnings of Western Science: The European Scientific Tradition in Philosophical, Religious, and Institutional Context, 600 B.C. to A.D. 1450.* Chicago, IL: The University of Chicago Press, 1992; 474 pp.

Lindberg, Evjue-Bascom Professor of the History of Science at the University of Wisconsin, provides here the first one-volume history of ancient and medieval science ever published. After introducing early views on the natural world and discussing the nature of science, Lindberg considers some of the methodological problems facing the science historian. He suggests that literacy was a necessary precursor to the rise of both philosophy and science. He traces the development of views on the cosmos from Homer, through to the time of the Middle Ages, and devotes entire chapters to the influence of Aristotle and the rise of science in Islam. He notes that the exclusion of supernatural explanations from scientific writings occurred as far back as Hippocrates, then again in twelfth century Europe. He also shows that a common twelfth century view was that God created, then let nature take its own uninterrupted course. He counters the popular notion that medieval scholars preferred logical reasoning over experimentation because they were willfully ignorant or dogmatic. Instead, Lindberg argues, that logic carried more weight because of a then popular mistrust of sense perceptions.

_____, and Ronald L. Numbers (eds.). *God and Nature: Historical Essays on the Encounter Between Christianity and Science.* Berkeley: University of California Press, 1986; 516 pp.
This volume grew out of an international conference on the historical relations of Christianity and science held at the University of Wisconsin in 1981. Sixteen historians provide eighteen chapters reviewing science/faith interactions from the early days of the Christian church, through the Middle Ages, and on into the present day. This is a scholarly work: each chapter contains numerous references, and a 27-page index makes the volume maximally useful. The authors address the question: Have science and Christianity been in conflict, or have they really been allies all along? The answer is that the interactions have been too complicated for any simplistic conclusion.

Livingstone, David N. *Darwin's Forgotten Defenders.* Grand Rapids, MI: Eerdmans, 1987; 210 pp.
This book provides the first systematic study of the responses of nineteenth and early twentieth century evangelical intellectuals to Darwinism. Livingstone demonstrates that rather than impeding acceptance of Darwin's theory of evolution, many evangelicals in both Britain and the United States came to accept and even defend the

theory. Christian defenders of Darwin were not only scientists, such as Asa Gray, James Dana, and William Dawson, but also clerics and theologians, such as George Frederick Wright, Charles Hodge, and B. B. Warfield. Livingstone effectively destroys the myth that "reactionary religion has always stood in the way of radical science."

Lubenow, Marvin L. *"From Fish to Gish": Morris and Gish Confront the Evolutionary Establishment.* San Diego, CA: CLP, 1983; 293 pp.
Lubenow is pastor of the First Baptist Church of Fort Collins, Colorado. He is a member of several creationist organizations and a frequent contributor to the creationist literature. This book recounts the exploits of Henry Morris and Duane Gish during their numerous debates with evolutionists at colleges, universities, schools, churches, and television and radio stations from 1972 to 1982. In most of these debates, the creationists "won." Selected debates are detailed in the main body of the book, while two appendices list one hundred and thirty-six debates by location and participants. Lubenow estimates that over five million people "have seen, heard, or read a creation-evolution debate involving Henry Morris and/or Duane Gish."

McIver, Tom. *Anti-Evolution: A Reader's Guide to the Writings Before and After Darwin.* Baltimore, MD: Johns Hopkins University Press, 1988; 385 pp.
Although he is not an antievolutionist himself, McIver, an anthropologist and research fellow with the National Endowment for the Humanities, collects antievolution books and pamphlets as a hobby. This volume provides 1,852 annotated references to many of the antievolution books and tracts from before the time of Darwin to 1988. He obtained many of the references from the Institute for Creation Research, which houses one of the largest collections of creationist literature in the world. The annotations are informative and are delightful reading in their own right. Written in a nonpolemical style, this book is a highly recommended bibliographic source for anyone interested in the history and nature of creationism.

May, Gerhard. *Creatio Ex Nihilo: The Doctrine of "Creation Out of Nothing" in Early Christian Thought.* Translated by A. S. Worrall. Edinburgh: T&T Clark, 1994; 197 pp.
May, a professor of theology at the Johannes Gutenberg University

in Mainz, traces the origin and development of the creation *ex nihilo* concept so popular among scientific creationists. In this view, God created the world and the rest of the universe out of nothing. May posits that the doctrine of creation was not debated among Christians until the second century. At that time they began to promote the "out of nothing" position as an expression of their belief that God was before all and above all and as a reaction against the Platonic view of preexistent matter and the gnostic view of emanationism. The Christian theologians Tatian, Theophilus of Antioch, and Ireneaus all promoted this view. Their position clashed, however, with those of earlier theologians, such as Justin and Clement of Alexandria, who taught that God used pre-existing matter to create. This is a useful, scholarly work that addresses the historical roots of a prominant tenet of conservative creationism.

Mayr, Ernst. *The Growth of Biological Thought: Diversity, Evolution, and Inheritance.* Cambridge, MA: The Belknap Press of Harvard University Press, 1982; 974 pp.
Few people are as qualified to write on the history of evolutionary thought as Ernst Mayr, emeritus professor of zoology at Harvard University and one of the chief architects of twentieth century views on evolution. Mayr emphasizes the development of some of the major ideas of modern biology, particularly those related to systematics, genetics, and evolution. In addition, Chapters 2 and 3 attempt to place biology in the context of other sciences and examine its intellectual heritage. The final chapter addresses the nature and philosophy of science and considers the notion of scientific progress. This is a substantial, well-referenced work.

_____. *One Long Argument: Charles Darwin and the Genesis of Modern Evolutionary Thought.* Cambridge, MA: Harvard University Press, 1991; 195 pp.
Unlike several other books by Mayr, this one is written for students and nonspecialists. After a brief introduction to Darwin himself, Mayr traces the history of evolutionary theory from Darwin's time to the present. He examines concepts like natural selection, speciation, and punctuated equilibrium, as well as the impact of Darwin's views on theology, philosophy, and diverse areas of science. He also discusses Darwin's shift from Christian faith to agnosticism. A glossary and long list of references make this an excellent reference

book on evolution and its history for the nonspecialist, a book written by a person who has played a large role in shaping modern evolutionary thought.

Miller, Jonathan. *Darwin for Beginners.* London: Writers and Readers Publishing Cooperative, 1982; 176 pp.
This little cartoon treatment of Charles Darwin and his ideas is one of a series of books on famous personages. It is entertaining, breezy, and humorous, yet intended to inform. Darwin's voyage on the *Beagle,* the development of his ideas, publication of *Origin of Species,* controversies after its publication, creationist objections, the emergence of neo-Darwinism, and the development of the theory of punctuated equilibrium in the twentieth century are all covered. As one might expect for a book of this style and format, events and ideas are significantly oversimplified and sometimes inaccurately reported. The reader who understands these limitations will, nevertheless, find this little volume fun to read.

Moore, James. *The Darwin Legend.* Grand Rapids, MI: Baker Books, 1994; 218 pp.
One of the myths enjoyed by evangelical Christians during much of the twentieth century has been that, on his deathbed, Charles Darwin gave up the theory of evolution in exchange for faith in the Bible. This story has been repeated countless times in scores of sermons, tracts, and articles since it was first published in 1915. Its originator, Elizabeth Reid Cotton—alias "Lady Hope"—a onetime temperance reformer from England, claimed to have witnessed Darwin's conversion during a visit with him shortly before he died. Although the story's primary assertion is fictitious—Darwin died an agnostic and never retreated from evolutionism—Moore presents evidence to suggest that Lady Hope did, in fact, interview the ailing evolutionist. This is a fascinating and engaging piece of history.

Moore, John A. *Science as a Way of Knowing: The Foundations of Modern Biology.* Cambridge, MA: Harvard University Press, 1993; 530 pp.
Moore, a biologist, examines the history of biological concepts from the Paleolithic period to the present. This book grew out of a series of eight essays published in *American Zoologist* between 1984 and 1990, part of a cooperative effort to encourage an understanding of the

roots of our knowledge as we seek to solve pressing environmental problems. Part I examines the history of human approaches to nature, from prehistoric times, through the era of Greek philosophy, early Judeo-Christian thought, the Renaissance, and pre-Darwinian views about the history of the earth and life. Part II recounts the origin and development of evolutionary thought against the backdrop of natural theology and Paley's argument from design, and also discusses evidence bearing on the history of life from an evolutionary perspective. Part III examines early views about heredity and cells, the emergence of Mendelian genetics, and twentieth-century advances in our understanding of DNA and genetic processes. Part IV addresses the "enigma of development," correctly highlighting this as a central challenge to the efforts of biologists. In a brief concluding essay, Moore reflects on the meaning of our "extraordinary" scientific understanding of many fundamental life processes. He writes that "Science deals not with the gods above but with the worlds below. It does not refute the gods; it merely ignores them in its explanations of the natural world." "[H]uman life," he concludes, "will always be constrained by the basic laws of nature, which the gods cannot annul."

Morris, Henry M. *A History of Modern Creationism.* Second edition. Santee, CA: Institute for Creation Research, 1993; 444 pp.
First published in 1983, this book documents the growth of the creationist movement from the point of view of its most prominent insider. Morris says that although he cannot claim the objectivity of a historian, he has "made every attempt to be factual." Indeed, the book is written in the first person, and the many facts presented are intertwined with autobiographical detail and editorial perspective. This is what gives this book its charm. For anyone seeking to better understand how creationists think and how they view the world, this book is an excellent place to begin.

Nebelsick, Harold P. *Circles of God: Theology and Science from the Greeks to Copernicus.* Dover, NH: Scottish Academic Press, 1985; 284 pp.
In a review of the history of theology and science, Nebelsick suggests that science and religion have served to prod one another along to new heights. Whereas religion provides science with a reason to explore, science could not reach beyond a rudimentary understanding of reality if it remained "captive to . . . religious ideology." If science does not

reach ahead, "religion is likely to remain caught in its own presuppositions and remain insipid as well." Nebelsick supports his case through a careful consideration of the religiously inspired efforts of Copernicus to defend the mistaken notion of circular planetary orbits. Kepler, he argues, was able to break through the theological impasse and thereby discover the elliptical shape of the orbits.

Nelkin, Dorothy. *The Creation Controversy: Science or Scripture in the Schools.* Boston, MA: Beacon, 1982; 242 pp.
Nelkin is a sociologist concerned over the influence of creationism in the public schools. Here she examines efforts by creationists at textbook censorship and also details events surrounding the 1981 "Scopes II" trial over the teaching of creationism in the public schools of Arkansas. She also discusses *Man: A Course of Study (MACOS)*, a naturalistic social science curriculum funded by the National Science Foundation and used with elementary school students during the early 1970's. According to Nelkin, the acclaimed curriculum ultimately failed because of clashes with the conservative religious community. Science and science education, she believes, must be able to operate free from nonscientific influences such as creationism if America is to keep pace in an increasingly competitive world.

Nelson, Paul (ed.). *The Creationist Writings of Byron C. Nelson.* Volume 5 of *Creationism in Twentieth-Century America,* Ronald L. Numbers, series editor. New York: Garland, 1995; 505 pp.
Byron C. Nelson (1893-1972) was a Lutheran pastor and an admiring disciple of George McCready Price. Early in his life, Nelson was a believer in Arnold Guyot's "day-age theory" of Genesis 1, but he began to defend a short chronology and flood geology after reading Price's *New Geology.* Along with a biographical sketch of Nelson, the following four of his books are reprinted in this volume: *"After Its Kind": The First and Last Word on Evolution* (1927), *The Deluge Story in Stone: A History of the Flood Theory of Geology* (1937), *A Catechism on Evolution* (1937), and *Before Abraham: Prehistoric Man in Biblical Light* (1948).

Numbers, Ronald L. *The Creationists.* New York: Alfred Knopf and Sons, 1992; 458 pp.
Numbers is a professor of the history of science and medicine at the University of Wisconsin. This comprehensive and well-referenced

volume won the Albert C. Outler Prize in Ecumenical Church History of the American Society of Church History in 1991. Although the early chapters highlight nineteenth century creationist views, the history of modern "creation science," from its relatively obscure beginnings in the early twentieth century to its virtual domination in fundamentalist circles today, provides this book's underlying theme. Creationist personalities and organizations receive excellent coverage. Numbers emphasizes American creationism but devotes one chapter to the history of the movement in Great Britain. Both creationists and evolutionists praise this work as an evenhanded, superbly written history of the creationist movement.

_____ (ed.). *Antievolutionism Before World War I.* Volume 1 of *Creationism in Twentieth-Century America,* Ronald L. Numbers, series editor. New York: Garland, 1995; 403 pp.
This collection of four works published before World War I provides glimpses into an interesting period of transition among creationists. As Numbers points out in the introduction, few thoughtful creationists between the time of Darwin and the early twentieth century were what we would call today "special creationists." Although they were not "creationists" in today's sense of the word, many Christians of the early twentieth century entertained serious doubts about Darwinian evolution. This skepticism was shared by mainline geneticists who had become convinced that natural selection was an insufficient cause of evolution. The four selections reprinted here—one by a German evolutionist and the other three by respected American clerics—reflect this skepticism of Darwinism.

_____ (ed.). *Creation-Evolution Debates.* Volume 2 of *Creationism in Twentieth-Century America,* Ronald L. Numbers, series editor. New York: Garland, 1995; 505 pp.
Debates have played a major role in the history of the creation/evolution controversy. Given their much-practiced art of rhetoric, creationists have often been the "winners" of these contests. This volume contains the texts of eight debates over creationism held between 1925 and 1937. Seven of the eight were between creationists and evolutionists, but one featured two creationists—William Bell Riley, who interpreted Genesis from a "day-age" framework, and Harry Rimmer, who argued for the "gap theory."

_____ (ed.). *The Antievolution Works of Arthur I. Brown.* Volume 3 of *Creationism in Twentieth-Century America,* Ronald L. Numbers, series editor. New York: Garland, 1995; 208 pp.
Arthur I. Brown (1875-1947) was one of the most prominant antievolutionists of the first half of the twentieth century. A surgeon by training, Brown was scientifically one of the best-informed creationists of his time. By 1925, requests for Brown's services as speaker led him to abandon his lucrative practice to devote all of his time to writing and lecturing on science and the Bible. He focused on examples of apparent design in nature as evidence for a creator. This collection features six pamphlets by Brown published during the 1920's.

_____ (ed.). *Selected Works of George McCready Price*. Volume 7 of *Creationism in Twentieth-Century America,* Ronald L. Numbers, series editor. New York: Garland, 1995; 489 pp.
Seventh-day Adventist educator George McCready Price (1870-1963) was the most influential creationist in America during the first part of the twentieth century. Indeed, the journal *Science* identified Price as "the principal scientific authority of the Fundamentalists." It was Price's "flood geology" that inspired Whitcomb and Morris' *The Genesis Flood* (1961), the book that ultimately led to development of the contemporary creationist movement. Numbers' biographical sketch of Price is followed by four of Price's works: *Illogical Geology: The Weakest Point in the Evolutionary Theory* (1906), *Q.E.D.: Or, New Light on the Doctrine of Creation* (1917), *The Phantom of Organic Evolution* (1924), and *Theories of Satanic Origin* (no date).

_____ (ed.). *The Early Writings of Harold W. Clark and Frank Lewis Marsh.* Volume 8 of *Creationism in Twentieth-Century America,* Ronald L. Numbers, series editor. New York: Garland, 1995; 531 pp.
Harold W. Clark (1891-1986) and Frank L. Marsh (1899-1992) were students of George McCready Price and among the first conservative creationists of the twentieth century to receive advanced degrees in science. In Clark's *The New Diluvialism* (1946), reprinted here, he outlined his views on flood geology, which differed significantly from those of Price. Most importantly, Clark accepted the predictable sequence of fossils in the geologic column, whereas Price did not. Clark accounted for this sequence through his "ecological zonation theory," which suggested that rising floodwaters wiped out successive

ecological life zones which are now preserved as the fossil sequence. Also reprinted here is Clark's *Back to Creationism* (1929). Marsh was more interested in biology than geology, most specifically in the amount of change in organisms since their creation. This collection includes Marsh's *Fundamental Biology* (1941), which details his views on biological change. Numbers introduces the volume with an informative biographical essay on both men and discusses their interactions with Price and with each other.

_____ (ed.). *Early Creationist Journals.* Volume 9 of *Creationism in Twentieth-Century America,* Ronald L. Numbers, series editor. New York: Garland, 1995; 629 pp.
Numbers introduces this volume with an essay on efforts by creationists to start their own journals. Early attempts were short-lived; creationists during the early twentieth century often published their articles in more generic fundamentalist publications like the *Bible Champion.* By 1937, however, after several false starts, a mimeographed journal, *The Creationist*, appeared. It was written and edited by a rancher, Dudley Joseph Whitney, who was an enthusiastic proponent of George McCready Price's flood geology. *The Creationist* made it only to the second volume before folding. In 1938 Price and his friends formed the Society for the Study of Creation, the Deluge, and Related Science, also known as the Deluge Geology Society. In 1941 it began publication of the professional-looking *Bulletin of the Deluge Society and Related Sciences* which survived for five volumes. In 1945 the Deluge Geology Society disbanded as a result of internal disputes over the age of the earth. The Natural Science Society was formed to take its place, and under its auspices a new journal, *Forum for the Correlation of Science with the Bible,* was launched; it survived for two volumes. This book contains the full text of *The Creationist* (1937-1938), *Bulletin of the Deluge Society and Related Sciences* (1941-1945), and *Forum for the Correlation of Science with the Bible* (1946-1948).

Osler, Margaret J. *Divine Will and the Mechanical Philosophy: Gassendi and Descartes on Contingency and Necessity in the Created World.* New York: Cambridge University Press, 1994; 284 pp.
Osler, of the University of Calgary, highlights the differing theological perspectives of two seventeenth century scholars, Gassendi and Descartes. The rationalist Descartes believed that God had initially

created a set of natural laws to which he was later bound. In this view, it is possible to have an *a priori* understanding of nature and eternal truths, because God created our minds in accordance with the rational structure of the world. By contrast, the voluntarist Gassendi believed in the contingency of nature, that God is bound by no natural law. God is bound only by the law of non-contradiction. Thus, we cannot reason our way to an understanding of nature; instead we must observe natural phenomena directly to know about them. This book provides important schoarly insights into the influence of Christian theology on the development of modern scientific thought.

Redondi, Pietro. *Galileo: Heretic.* Princeton, NJ: Princeton University Press, 1987; 356 pp.
Redondi, a French science historian, provides a fascinating revisionist view of Galileo's trouble with the Catholic church during the seventeenth century. The standard view is that Galileo was brought to trial because of his Copernican heliocentric views. According to Redondi, this is only partly true. Based on new evidence, Redondi argues that Galileo's troubles were based more in his view that the world is composed of invisible atoms. This notion, according to Galileo's Jesuit foes, discredited the church's position that the Sacrament contained the literal presence of Christ. According to Redondi, the charge of Copernicanism for which Galileo was tried was merely a smoke screen for the real issue of atomism. Pope Urban VIII, a friend of Galileo, orchestrated the trial over the lesser charge of Copernicanism, thereby placating the Jesuits and saving Galileo's life.

Richards, Robert J. *Darwin and the Emergence of Evolutionary Theories of Mind and Behavior.* Chicago, IL: The University of Chicago Press, 1987; 688 pp.
Richards, who teaches history, philosophy, and behavioral science at the University of Chicago, traces the theories of mind and behavior over the past two centuries. The perspectives of Charles Darwin provide a primary focus for Richards, as well as the views of Darwin's predecessors and contemporaries and those Darwin himself would later influence. Richards shows that although many modern scientists operate on the assumption of mechanistic materialism, many early evolutionists had developed "carefully worked out metaphysics . . . completely opposed to mechanistic materialism." Richards also deals with reasons for Darwin's delay of publication of his *Origin of Species* and

discusses the relationship between ethics and evolutionary theory. This is a carefully researched, informative study.

_____. *The Meaning of Evolution: The Morphological Construction and Ideological Reconstruction of Darwin's Theory.* Chicago, IL: The University of Chicago Press, 1992; 205 pp.
Richards provides a carefully researched analysis of the background and nature of Darwin's evolutionary theory. Richards describes the changes that occurred in the meaning of the term "evolution." During the seventeenth and eighteenth centuries evolution meant simply individual embryological development. By the early nineteenth century meaning of the term had shifted to refer to the supposed series of lower forms of life through which an embryo progressed on its way to a more perfected form. Eventually, evolution came to refer to species change over generations. Assessments by Ernst Mayr, Stephen Gould, Peter Bowler, and other recent writers to the contrary, Richards asserts that Darwin was a child of his time—a Lamarckian recapitulationist and progressivist. Twentieth century evolutionists who think otherwise, he says, are simply reading their own ideas back into Darwin's writings in an effort to receive the honor of Darwin's posthumous blessing.

Roberts, Jon H. *Darwinism and the Divine in America: Protestant Intellectuals and Organic Evolution, 1859-1900.* Madison: University of Wisconsin Press, 1988; 242 pp.
In this careful review of primary sources, Roberts evaluates the responses of American Protestant intellectuals to Darwin's theory of evolution. The initial reaction to *Origin of Species,* Roberts contends, revolved around its perceived lack of scientific merit. After 1875, however, Protestants focused more on the theological implications of Darwinism. During this time, some Protestant thinkers began to look for ways to accommodate their theology to Darwinism rather than seeking to denounce it. Roberts believes that it was a general Protestant enthusiasm for science, developed before the publication of Darwin's book, that eventually led to the devaluation of Christian theology. Protestants seemed unable to balance theological and scientific perspectives, ultimately caving in to scientific dogma and authority. For this insightful and well-written work, the author won the Brewer Prize of the American Society of Church History.

Rossi, Paolo. *The Dark Abyss of Time.* Chicago, IL: The University of Chicago Press, 1986; 388 pp.
This scholarly work was written to illuminate the philosophical discussions surrounding the prominant seventeenth century Italian philosopher Vico, author of *The New Science*. Many of these discussions centered on earth and human history in relation to the Bible. Interestingly, some of these discussions were very similar to those held today. For example, in his treatment of seventeenth century geology, Rossi points out that the age of the earth was frequently debated. Just as today, some maintained that the earth was only a few thousand years old, while others believed that it was millions of years old. Rossi's book makes an important contribution to our understanding of seventeenth century thought.

Rusch, Wilbert H., Sr., and John W. Klotz. *Did Charles Darwin Become a Christian?* Norcross, GA: Creation Research Society Books, 1988; 38 pp.
Rusch is an emeritus professor of science and mathematics at Concordia College, Michigan, and Klotz is professor of natural science at Concordia Senior College, Indiana. Both are prominent creationist authors. Here they evaluate the commonly told tale that Charles Darwin reconverted to Christianity in 1882, close to the time of his death. The authors trace this story back to a mysterious "Lady Hope," whose true identity they never ascertain. Using the testimony of Darwin's daughter and of Darwin himself, they argue that "Lady Hope's" story is completely false and that no such person ever saw Darwin. Based on the evidence they discuss and contrary to the popular rumor, the authors conclude that Darwin died an agnostic.

Trollinger, William Vance (ed.). *The Antievolution Pamphlets of William Bell Riley*. Volume 4 of *Creationism in Twentieth-Century America,* Ronald L. Numbers, series editor. New York: Garland, 1995; 221 pp.
Trollinger provides a biographical sketch of William Bell Riley (1861-1947), a Baptist pastor and antievolutionist. Riley was a leader in the fundamentalist movement between the two world wars. His World's Christian Fundamentals Association, founded in 1919 to stop the teaching of evolution, was involved in selecting textbooks appropriate for Christian schoolchildren. Like many other conservative Christians of his day, Riley defended the "day-age theory" which

allowed for long ages. This volume contains nine of Riley's anti-evolution pamphlets.

Urbach, Peter. *Francis Bacon's Philosophy of Science.* LaSalle, IL: Open Court, 1987; 209 pp.
Francis Bacon is often credited with developing "the scientific method." While most scientists and philosophers today eschew a purely Baconian approach to the natural world, scientific creationists often define science in Baconian terms. This book examines Bacon's approach to knowledge, often thought of as simply the inductive method. Urbach suggests that a better term to describe Bacon's approach is the hypothetico-inductive method. The hypothetico-inductive method leads from data to middle axioms; the middle axioms, in turn, are used to ascertain other relationships in the data; these other relationships, in turn, allow for the development of further axioms—and so on. Bacon believed that humans should come to data with as few preconceived ideas as possible. This is accomplished, he believed, by dismissing five common mindsets: (1) the assumption that there is more order and regularity than there really is, (2) that unusual events should be ignored in the development of axioms, (3) that reason governs words, (4) that the study of phenomena begins with previously formed rational explanations, and (5) that less interesting observations are insignificant. Urbach also discusses the role of experiment in science in view of Bacon's ideals.

Webb, George E. *The Evolution Controversy in America.* Lexington: University of Kentucky Press, 1994; 297 pp.
Webb, a professor of history at Tennessee Technological University, examines the responses of Americans to evolutionary concepts, looking especially at the effect of these responses on public school education in the sciences. Two introductory chapters briefly outline the reaction of American society to Darwinism, consider the development of neo-Lamarckism and neo-Darwinism, and examine how evolutionary theory has influenced American religion and society. Remaining chapters focus on the impact of the creation/evolution controversy on public education. Many science teachers and parents, he says, believe that science education is best served when students learn about nature within an evolutionary context. By contrast, their fundamentalist Christian opponents believe that evolutionary theory leads students to accept an atheistic outlook of the world. In this context

Webb traces the various methods used by antievolutionists to effect changes in the public school curriculum. He is clearly opposed to the creation science movement and is particularly concerned about the effect this movement is having on science education in America.

Wybrow, Cameron (ed.). *Creation, Nature, and Political Order in the Philosophy of Michael Foster (1903-1959): The Classic Mind Articles and Others, with Modern Critical Essays.* Lewiston, NY: Mellen, 1992; 347 pp.
Michael Foster, a tutor at Christ College, Oxford, was a prominent British philosopher interested in the interplay between theology and science. In an introductory essay, Wybrow recounts the story of Foster's difficult life, intellectual influences, scholarly endeavors, and eventual suicide. Foster was committed to the notion that modern science actually stands indebted to Christian theology. Christian theology holds that God's creation is contingent, not necessary, a view that promotes seeing the world empirically, like modern science, rather than from the classical a priori perspective. This volume includes three papers by Foster published during the 1930's in the journal *Mind*, as well as four additional articles and an annotated bibliography of Foster's works. Also included are eight essays written in response to Foster's ideas.

Young, David. *The Discovery of Evolution.* London: Natural History Museum Publications, 1992; 256 pp.
Beginning with the observations of naturalists in the seventeenth century, Young uses historical narrative to explain what the theory of evolution is about and how it came to be. He first provides a broad sketch of evolution by natural selection and the types of evidence this theory seeks to explain. He then traces the growing interest in natural diversity from John Ray, through Carl von Linne, to the time of Charles Darwin, and on into today. He discusses early attempts to sort out the history of the earth and past life by Comte de Buffon, James Hutton, Georges Cuvier, Charles Lyell, and others. The story of Charles Darwin and the development and reception of his theory of evolution by natural selection is recounted. Early work in genetics by Gregor Mendel and others is described, as is the synthesis of ideas in paleontology, genetics, and evolutionary theory during the early twentieth century. Recent developments in evolutionary theory are also discussed. Toward the end of the book is an "Evolutionary

Who's Who," containing the names and brief descriptions of people who contributed to the development of evolutionary theory. Suggested reading is also provided, along with a list of references. One of the nicest features of this book is the fine set of illustrations, many from historical sources.

Chapter 4

PHILOSOPHICAL, THEOLOGICAL, AND GENERAL REFERENCES

The creation/evolution controversy ultimately centers around questions of philosophy and theology: Why do we perceive reality in the ways that we do? Could we interpret reality differently from how we interpret it? Is there a God? If God exists, what is he or she like and how does he or she interact with the universe? If God does not exist, where did the universe come from? What is the meaning of existence? These are, of course, very big questions but there is no shortage of answers. The references in this chapter supply some of these answers and consider other general questions related to the origin and history of the universe. This chapter also includes references that address a mixture of topics related to the creation/evolution discussion.

THEIST REFERENCES

Amos, Gary T. *Defending the Declaration: How the Bible and Christianity Influenced the Writing of the Declaration of Independence.* Brentwood, TN: Wolgemuth and Hyatt, 1989; 235 pp.

This book attacks the notion that the Declaration of Independence is rooted in deistic philosophy rather than biblical Christianity. Amos claims that the Declaration's use of the phrase "laws of nature and of nature's God" was not at all intended to be interpreted from a deistic perspective. He quotes other seventeenth and eighteenth century documents that use similar phraseology but which provide greater elaboration and clearly demonstrate Christian worldviews. These quotations, he says, demonstrate that the Declaration's parallel wording should also be interpreted within a Christian and creationist framework. Amos goes on to argue that Genesis 1 provides the basis for inalienable rights in human society.

Aviezer, Nathan. *In the Beginning: Biblical Creation and Science.* Hoboken, NJ: KTAV, 1990; 138 pp.
Aviezer is a professor of physics at Bar-Ilan University, Tel Aviv. In this examination of the relation between science and scripture, Aviezer makes frequent reference to the writings of medieval Jewish scholars of the Torah, including Radak, Ramban, and Rashi. He interprets creation as having occurred over a fifteen-billion-year period, with each of the six days lasting two and one half billion years. According to Aviezer, the expression "Let there be light" in Genesis 1:3 refers to the Big Bang. He accepts the theory that the moon formed as the result of a collision between some other planetary body and the earth. He also subscribes to Darwinian evolution, believing that the fossil record provides ample evidence for this view. This book provides an interesting Jewish perspective on the creation/evolution controversy, a conflict that, at least in recent years, has more frequently emerged as a point of discussion in the Christian community.

Barnes, Thomas G. *Science and Biblical Faith: A Science Documentary.* Kansas City, MO: Creation Research Society Books, 1993; 191 pp.
Barnes is a creationist and an applied physicist. He was an early member of the Creation Research Society. This book contains a series of biographies of outstanding scientists who were Christians: Lord Kelvin, Louis Pasteur, Isaac Newton, Johannes Kepler, Michael Faraday, James Maxwell, Ernest Rutherford, Lord Rayleigh, Joseph Thomson, and W. R. Thompson. This list is extended by Barnes to include the stories of several modern creation scientists: Walter Lammerts, Harold Slusher, Robert Gentry, and Donald DeYoung. In addition to these biographical pieces, Barnes provides a critique of modern physics, which he believes is too mathematically complicated. He believes that classical physics provides superior solutions to physical problems. He reviews some of his own work supportive of a young earth model and discusses the results of his research into electromagnetic feedback and light propagation. He posits that creationism, unlike evolutionism, is consistent with the laws of thermodynamics, and he scorns the modern view that order can be produced naturally from chaos and disorder.

Berry, R. J. (ed.). *Real Science, Real Faith.* East Sussex, England: Monarch, 1991; 218 pp.

This book is a collection of essays by sixteen leading scientists who believe that an appropriate and thoughtful integration of science and faith is possible. Contributors include the director of Kew Gardens, director-general of The Meteorological Society, secretary of the International Whaling Commission, and nine professors, one an evolutionary biologist. The authors express a variety of religious beliefs and sentiments: for example, that God continually acts to uphold the creation; that Christians have a responsibility to care for the earth; that science and Christianity approach the same truth from different points of view; that faith and science are complementary, not contradictory; and that we would benefit by accepting the biblical view of a psycho-physical unity.

Bessinger, Donivan. *Religion Confronting Science: And There Was Light.* Greenville, SC: Orchard Park, 1991.
Bessinger, an Episcopal surgeon, writes in an attempt to find harmony between science and the Christian faith. He believes that when both perspectives are properly understood, there is no conflict between them. He describes a progression of worldviews, from ancient mythology, to Medieval alchemy, to the modern scientific cosmology. He believes that the modern scientific worldview is devoid of the meaning found in earlier mythological cosmologies. Consequently, he suggests that today's challenge is to develop a "grand synthesis" that incorporates both scientific reason and spiritual meaning. Bessinger is critical of biblical literalists because he believes they miss the deeper meanings of the biblical text. Only by reading scripture with an intuitive, "mythic eye," he suggests, will we find its true meaning in relation to science. He believes that evolution is responsible for shaping life and that natural selection may interfere with our ability to detect design, except at the basic level of life principles. Quantum mechanics, he suggests, points toward a reality that transcends science.

Binns, Emily. *The World as Creation: Creation in Christ in an Evolutionary World View.* Wilmington, DE: Michael Glazier, 1990; 104 pp.
Binns is a professor of systematic theology at Villanova University. In this book she provides a worldview consistent with the writings of Teilhard de Chardin, whom she studies and admires. She first provides a biblical and historical overview of the problem of creation, followed

by a discourse in which she identifies humans as co-creators with God. She then examines the challenges presented to Christians in the context of environmental degradation, world hunger, and poverty. She concludes with a chapter on hope in the context of Christology. To Binns, Christ is more than a historical figure; he is also a "radically transformed, spiritual presence filling the cosmos." Along with Teilhard, she envisions evolution as leading to progressively higher levels of complexity.

Bird, W. R. *The Origin of Species Revisited I.* New York: Philosophical Library, 1989; 551 pp.
Bird, a Yale Law School graduate and former editor of the *Yale Law Journal,* has been active in the scientific creationism movement since the late 1970's. This is the first of a two-volume work on the creation/evolution controversy. In Part I of this volume, Bird defines "theories of abrupt appearance" (his term for a variety of creationist views), evolution, and the religious meanings of both; Part II provides arguments for abrupt appearance and against evolution; Part III deals with the origin of life; and Part IV is concerned with the origin of the universe. Within this framework, he deals with a diverse, but well-organized, collection of topics including systematics and typology, panspermia, vitalism, neo-Darwinism, saltational change, macromutations, neo-Lamarckism, and theistic evolutionism. He attempts to use the writings of evolutionists themselves to discredit evolutionary theory. This book provides a comprehensive, "lawyerly" defense of scientific creationism.

Bixler, R. Russell. *Earth, Fire and Sea: The Untold Story of Creation.* Pittsburgh, PA: Baldwin Manor, 1986; 214 pp.
Bixler is president of Cornerstone TeleVision of Pennsylvania, a creationist, and a believer in the inerrancy of the Bible. He believes that God's laws today are the same laws he used to create. He attempts to show that the view that God created out of nothing—creation *ex nihilo*—is unbiblical. Life was created in the relatively recent past, he believes, but matter and the universe are very old. Thus, for Bixler, light from stars and galaxies millions of light years away, as well as the ages of rocks determined by radiometric dating, are not problems. He posits that rapid cooling of the surface of the earth was responsible for the formation of pleochroic halos in ancient rocks. He opposes the gap theory, the day-age theory, and

the view that God revealed the creation in six days to the writer of Genesis.

Blocher, Henri. *In the Beginning: The Opening Chapters of Genesis.* Downer's Grove, IL: InterVarsity, 1984; 240 pp.
Blocher is a French evangelical scholar. This book is a translation (by D. G. Preston) of his book *Revelation des Origines: Le Debut de la Genese* (1979). He begins by discussing his methodology and his basic assumption that the biblical text is the word of God. He also realizes that this word was written by humans, writing in different styles and from different perspectives. He believes that one should read the text in relation to the rest of scripture. He takes Genesis 1 to be a "hymn of praise" and thinks that it is a distortion to interpret this text according to any scientific scheme. Blocher is critical of the "documentary hypothesis," the view that Genesis 1 and 2 were compiled from several sources. Although he argues that Genesis 2 and 3 should be taken as historically informative, he does not believe that theistic evolution is necessarily contraindicated by the text. He is not sure where Adam should appear in relation to the geologic column but believes that, ultimately, this problem will be resolved.

Boys, Don. *Evolution: Fact, Fraud or Faith.* Largo, FL: Freedom, 1994; 353 pp.
Don Boys, a writer, a former member of the Indiana House of Representatives, and a Baptist evangelist, received a Ph.D. from Heritage Baptist University. In this book he declares war on evolutionists, a war in which he says he is not a "conscientious objector." He says he is "willing to bare-knuckle it with any evolutionist." According to Boys, God created the entire universe in six literal days less than ten thousand years ago. He says it surprises him that "informed people don't fall to the floor, gasping and holding their sides with raucous laughter" when they hear evolutionary "drivel." The book highlights the common arguments used by creationists to support young earth beliefs and to counter evolutionary theory. Among other unbecoming epithets, Boys calls evolutionists venomous, uncivil, pathetic, venal, unscholarly skunks. An extensive list of references concludes this diatribe by an angry and bitter creationist.

Bube, Richard H. *Putting It All Together: Seven Patterns for Relating Science and Christian Faith.* Lanham, MD: University Press of

America, 1995; 213 pp.
Bube, emeritus professor of materials science and electrical engineering at Stanford University, is well known for his work in the realm of science and faith. Here he reviews a large body of literature containing a wide range of perspectives on this topic. He distills these perspectives down to seven basic patterns of thought that he then evaluates. Bube is concerned that readers understand what constitutes authentic science and authentic Christian theology. Only in this way will they be able to evaluate the various perspectives appropriately and avoid pseudoscience, on one hand, and pseudotheology, on the other. Bube is critical of scientism, inerrantist interpretations of the Bible, and New Age perspectives. He believes that the best approach to science/faith issues is one of complementarity—that the best picture of nature results when insights from two or more models of reality are integrated.

Buell, Jon, and Virginia Hearn (eds.). *Darwinism: Science or Philosophy?* Richardson, TX: Foundation for Thought and Ethics, 1994; 229 pp.
This volume contains the proceedings of a symposium held at Southern Methodist University in 1992. Contributors of the twenty-five chapters include philosophers and scientists, both theistic and naturalistic in outlook. Phillip Johnson begins the set of essays by distinguishing methodological naturalism, a feature of the most commonly accepted form of the scientific method, from metaphysical naturalism, the view that nothing beyond the purview of science is real. He claims that metaphysical naturalism is a necessary component of Darwinism. Michael Ruse, Leslie Johnson, and John Morrow each defend the position of metaphysical naturalism and Frederick Grinnell and Arthur Shapiro argue that science and religion deal with different realities. By contrast, essays by Stephen Meyer, Michael Behe, William Dembski, David Wilcox, and Peter van Inwagen support the notion that science should be open to both natural and supernatural realities as it seeks to interpret the universe.

Burke, Derek (ed.). *Creation and Evolution.* Leicester, England: Inter-Varsity, 1985; 288 pp.
This is a debate in print over the creation/evolution issue. The seven writers, holding views that dramatically differ from one another, nevertheless all subscribe to the following statement: “In the beginning

God created the heavens and the earth (Gen. 1:1). God's creation is a sovereign act of his power, love, and wisdom and depends on no other being or created substance. It was *ex nihilo*." Contributors range from Duane Gish, who discounts the notion that God would have used the "wasteful, inefficient, cruel method" of evolution to create, to R. J. Berry, who defends the evolutionary process and the existence of transitional forms. Other topics addressed include flood geology, radiometric dating, apparent age theories, the fossil record, and philosophical issues surrounding the creation/evolution debate.

Camp, Ashby L. *The Myth of Natural Origins: How Science Points to Divine Creation.* Tempe, AZ: Ktisis, 1994; 133 pp.
Camp pastors at University Church of Christ in Tempe, Arizona. He graduated with a degree from Harding University School of Religion and earlier worked as an attorney after earning a law degree from Duke University. He wrote this book to provide his daughter with an alternative to the "evolution propaganda flooding our society." He provides extensive quotations from previous authors that support his "critique of the naturalist's theory of origins." Camp addresses four main topics: the origin of the cosmos, the origin of life, the diversification of life, and the origin of humans.

Carvin, W. P. *Creation and Scientific Explanation.* Edinburgh: Scottish Academic Press, 1988; 106 pp.
Carvin, a Baptist minister and professor of religion and philosophy, received undergraduate training in math and physics. In this book, he examines the relationship between scientific cosmologies and the Christian doctrine of creation. Carvin believes that science and religion ask similar cosmological questions: Where did things come from? Why are things the way they are? The answers from the two different perspectives, he feels, may inform one another. He believes that Christians should talk about the universe in the context of our current cosmologies. He believes that the Christian faith can provide powerful and useful, although unprovable, answers to the ultimate cosmological questions.

Chittick, Donald E. *The Controversy: Roots of the Creation-Evolution Conflict.* Portland, OR: Multnomah, 1984; 280 pp.
Chittick is a chemist and educator. He wrote this book to help people deal with the creation/evolution debate. He believes that creationism

and evolutionism represent two antithetical worldviews. While he agrees that it is not a science textbook, he argues that the Bible provides a framework for the interpretation of scientific facts and that it contains accurate information about earth history. He rejects the findings of radiometric isotope dating, given that the initial isotope conditions cannot be known and that the isotope decay rates may not have remained constant. The days of Genesis 1, he says, were literal 24-hour days. He believes that it is difficut to develop an accurate history of the past due to mitigating factors like the fall, the flood, time, and cultural conditioning. The creation/evolution controversy, he contests, is not over facts, but over the interpretation of facts.

Clark, Gordon H. *The Philosophy of Science and Belief in God.* Jefferson, MD: Trinity Foundation, 1987; 140 pp.
Clark, a philosopher, is interested in the relationship between scientific theory and belief in God. He believes that attacks on the concept of God by scientists are naive and misdirected. He uses the process of motion to make his point. Motion is central to science. In essence, science is about attempting to understand processes of motion in nature; however, from the time of the Greeks to the present, no one has been able to explain motion beyond a superficial level. If, then, science has ultimately failed to explain this most fundamental of processes, science cannot be a source of truth; thus, science has no basis for denying the existence of God or his activity in the world. To Clark, science "is a knowledge of what to do in a laboratory. But a knowledge of nature, No." For him, ultimate value is derived not through science but through philosophy and theology.

Cochrane, Charles C. *The Gospel According to Genesis.* Grand Rapids, MI: Eerdman's, 1984; 88 pp.
Cochrane is a Presbyterian pastor. He posits that the real beginning of the biblical story occurs with Abraham in Genesis 12—chapters 1-11 serve as a prologue to the central salvation story in scripture. Genesis 1-11 reveal the identity of the creator and the created, but these chapters should not be taken as history or science. Instead, they consist of parables—truth presented in the form of story. In addition to exploring the primary theological considerations of the Genesis text, Cochrane also discusses biblical chronology, differences between the creation stories of Genesis 1 and 2, the order of creation, and other commonly discussed issues. At the end of the book Cochrane warns

against dogmatism on the part of either creationists or evolutionists and confesses that he finds "the evolutionary premise . . . quite inoffensive as an explanation for change."

Cogyne, George V., Karl Schmitz-Moorman, and Christoph Wassermann (eds.). *Studies in Science & Theology 1994: Origins, Time and Complexity (Part II).* Geneva, Switzerland: Labor Et Fides, 1994; 318 pp.
This book contains 46 papers by participants at the Fourth European Conference on Science and Theology, held in Italy in 1992. The conference was sponsored by the Vatican Observatory of the Vatican City State. Attendees representing both the sciences and humanities came from Europe, Africa, and North America. They considered a wide range of issues related to origins, time, and complexity. Opinions expressed in this volume represent a diversity of perspectives. The papers provide insights into the many ways Christians of various traditions deal with issues of science and faith.

Committee for Integrity in Science Education. *Teaching Science in a Climate of Controversy.* Ipswich, MA: American Scientific Affiliation, 1986; 64 pp.
The American Scientific Affiliation consists of a group of scientists who are Christians "committed to understanding the relationship of science to the Christian faith." This small book was designed to help public school science teachers and students sort through the maze of ideas related to origins and the history of life. It was published to provide balance to viewpoints expressed in *Science and Creationism: A View from the National Academy of Sciences* (1984), a booklet with a similar goal but written from a naturalistic perspective. The first section examines how to cope with the creation/evolution controversy and discusses blunders by overly zealous evolutionists and creationists in the past. The second section addresses the origin of the universe, the origin of life, the origin of animals, and the origin of humans. The book supports the Big Bang theory for the origin of the universe, is skeptical about the chemical evolution of life, suggests that microevolutionary processes do not adequately account for the origin of higher taxa, and remains noncommittal about human origins. The third section is addressed to teachers, urging them to "uphold the standards of scientific integrity by showing students how to arrive at conclusions based on valid evidence, and by teaching with openness."

An appendix recommends other sources for the teacher, and an addendum includes a series of classroom exercises on fossils, evolutionary relationships, and critical thinking.

Corey, Michael A. *The Natural History of Creation: Biblical Evolutionism and the Return of Natural Theology*. New York: University Press of America, 1995; 446 pp.
Corey claims in his preface to have previously proved, using the strong anthropic principle, the existence of a designer, one who correctly set all physical constants at the beginning. In this book Corey seeks to show that evolutionary theories about the history of the universe since the Big Bang correspond to the biblical creation story. For example, he believes that moral evil was necessary for human advancement. Thus, the fall into sin described in Genesis 3 was an important event in the moral improvement of humans which allowed them to be truly free. Corey notes that all scholarly disciplines explore the same reality, but that conflict exists between philosophical theology and science. He suggests that to remedy this problem, all scientists should obtain training in philosophy.

Cottrell, Jack. *What the Bible Says About God the Creator*. Joplin, MO: College, 1983; 518 pp.
This is the first of three books in a series on the doctrine of God, this one focusing on God as creator. Unlike many theologians who see God's redemption as primary to the Christian faith, Cottrell believes that God's creatorship is primary and that a proper understanding of that creatorship derives from the testimony of scripture. He believes that the Bible is inspired, infallible, and understandable, and that it is the most important source of knowledge about God. Cottrell discusses what he calls "pagan alternatives to creation," noting that the Christian view "stands out in unique contrast" to these other views. He believes that the doctrine of creation consists of three central concepts: that God created out of nothing, that creation was an act of freedom, and the creation includes both the physical and spiritual realms. He sees God as unique in that he is an infinite "Uncreated Spirit" unlimited by space and time.

Creager, John A. *Theodynamics: Neochristian Perspectives for the Modern World*. Lanham, MD: University Press of America, 1994; 452 pp.

Theodynamics, according to Creager—who wishes to make theology relevant in the modern world—is a view of God consistent with the notion of "the world as an organic process." According to Creager, God is continually present in the world, though that presence cannot be identified with any particular entity or process. Creager denies the Christian concept of the divine Trinity but accepts the Greek dualism of body and soul. He believes that sin and pain constitute a step in the evolution of humans toward a conscious participation with God. He considers the stories of creation and the fall of Adam and Eve to be myths in the same genre as the creation myths of pagan cultures.

Culp, G. Richard. *Remember Thy Creator.* Middlebury, IN: Historic Christian Publications East, 1990; 207 pp.
Culp is an osteopathic physician who converted to creationism during his days as a university student. He recounts many of his graduate student experiences in the book. He counters evolutionary theory through discussions of things like fossil hominids and pleochroic halos. He provides a detailed description of the human eye as an example of a designed structure and reviews the functions of the tonsils, appendix, and thymus, all once considered to be human vestigial organs. The book is a testament of the author's commitment to conservative creationism.

Durant, John (ed.). *Darwinism and Divinity.* New York: Basil Blackwell, 1985; 210 pp.
Seven contributors provide insight into the religious implications of Darwin's *Origin of Species.* Chapters include discussions of Darwin's influence on theology, Darwin's own religious journey, the accommodation of liberal Protestant theology with evolutionary theory, the use of natural selection as a mode of divine creation and incarnation, the effects of religion on human biology, the religious beliefs of atheists and agnostics, and reasons why scientific creationism has become influential. This is an interesting, scholarly, interdisciplinary work that will broaden the reader's understanding of the impact of Darwinism in the nineteenth and twentieth centuries.

Ecker, Ronald L. *The Evolutionary Tales.* Palatka, FL: North Bridge Books, 1993; 212 pp.
In verse reminiscent of Chaucer's *The Canterbury Tales,* the reader listens in on ten travellers heading to a creationist seminar in Dayton,

Tennessee, site of the 1925 Scopes "monkey" trial. The travellers include Astronomer, Biochemist, Biologist, Cosmologist, Geologist, Paleoanthropologist, Paleontologist, Philosopher, Physicist, and Scholar. During the trip, each traveller makes a case for the truth of evolution and the falsity of creationist claims. Ecker, through his protagonists, sees no conflict between the theory of evolution and the notion that God created life in the beginning. He believes, however, that the Genesis creation stories were derived from other Near Eastern creation myths and that young earth creation scientists are on the wrong track. He considers science and religion to be separate but important human endeavors. An extensive bibliography makes this novel little book a useful reference.

Faid, Robert W. *A Scientific Approach to Christianity.* Green Forest, AR: New Leaf, 1991; 196 pp.
Faid is a nuclear power industry consultant. He believes that the Bible was written in mathematical code in both Hebrew and Greek. He calls the study of this coded message "theomatics." The interlocking code of scripture is so impressive, says Faid, that God clearly "dictated it word for word." He believes that theomatics proves the historicity of Jesus and his resurrection, the reality of heaven and hell, and that it refutes the theory of evolution. He is critical of the Big Bang theory and claims that "Not one shred of hard evidence has come to light to prove that any single member of any genus has evolved from or into another genus." The sun, he believes, was created on the first day rather than the fourth day of creation week. The book contains a short bibliography.

Ferguson, Kitty. *The Fire in the Equations: Science, Religion, and the Search for God.* Grand Rapids, MI: Eerdmans, 1995; 320 pp.
Ferguson, a singer, conductor, and graduate of Juilliard School of Music, took up a career as a science writer after completing a stint at Cambridge University. In this unusual book she attempts to address the question of the existence of God by examining interactions among science, theology, and philosophy. She accomplishes this in a literary style somewhat foreign to readers accustomed to writings on topics of this nature. Whether discussing the strengths and limitations of science, evolutionary theory, or cosmology, Ferguson repeatedly concludes that science can neither prove nor disprove the existence of God. She believes that whether one considers God or

"Mathematical and Logical Consistency" as the cause for existence is ultimately a matter of faith.

Ferre, Frederick (ed.). *Concepts of Nature and God: Resources for College and University Teaching.* Athens: Department of Philosophy, University of Georgia, 1989; 258 pp.
The book resulted from The Summer Institute on Concepts of Nature and God held in 1987 at the University of Georgia. Participants discussed and pondered the question of what can be known about God by studying the universe. There are three main sections to the book: (1) "Workshop on Ancient & Medieval Thought," (2) "Workshop on Modern Thought," and (3) "Workshop on Contemporary Thought." The text consists primarily of annotated bibliographies, although there is some general discussion. This book will be useful to someone doing research on the topic of natural theology.

Feyerabend, Henry. *God's World.* London, Ontario: Arts International, 1986; 100 pp.
Feyerabend is a televangelist and the director of Adventist Radio Television Services of Arts International. This book consists of a series of seven sermons first presented on television. Feyerabend believes that science and the Bible are allies, and that "Nature and the Bible are two books written by the same Author." He claims that many modern scientific and technological feats were outclassed by biblical events: Christ healing the blind and the deaf, the preparation of a fish to carry Jonah, the floatation of a steel axehead, the transport of Philip through the air, and so on. He denies evolution and claims that belief in an earth millions of years old is necessary because scientists assume evolution, for which "time is the great hero." The earth and life, he believes, were created six thousand years ago with the appearance of age. The facts of astronomy, botany, zoology, oceanography, anthropology, and ornithology, he says, all testify to the creator God of the Bible who also wants to save those he has created.

Fischer, Dick. *The Origins Solution: An Answer in the Creation-Evolution Debate.* Lima, OH: Fairway Press, 1996; 382 pp.
Upon his return from Vietnam, where he flew numerous combat missions, Fischer earned his master's degree in theology. He is critical of the "warfare" metaphor often used to describe the relationship between science and religion. He believes that misunderstandings develop in

part because of inaccurate transcriptions of the Bible by scribes and because of inappropriate translations of Hebrew words into English. Fischer discounts creation science because of its weak science and faulty biblical interpretation. He believes that creation scientists discredit the Bible when they insist on things such as a young earth. Nonetheless, Fischer believes that Adam was a historical figure who lived about 5000 B.C., and that he was the first human to establish a covenant with God; moreover, life expectancy decreased when Adam's descendants interbred with their ungodly neighbors, and the flood of Noah, a local event, occurred around 3000 B.C.

Fischer, Robert B. *God Did It, but How?* La Mirada, CA: Cal Media, 1981; 113 pp.
Fischer, a chemist, is academic vice president at Biola College and a former dean of natural sciences and mathematics at California State University. He is particularly concerned over the frequent confusion between scientific and philosophical issues in the creation/evolution controversy. This confusion, he believes, is a major source of conflict. He posits that the biblical God is capable of operating either inside ("naturally") or outside ("supernaturally") the usual mechanisms of nature to achieve his purposes. Fischer carefully defines the concepts he uses, thus allowing the reader to understand his argument at each step of reasoning. He develops a model for Christian theism that sees God as both transcendent and immanent in nature and as both the creator and sustainer of the universe. In the final chapter Fischer shows that presuppositions used by members of the scientific community are analogous to those used by the theological community.

Fix, William R. *The Bone Peddlers: Selling Evolution.* New York: Macmillan, 1984; 337 pp.
The contemporary creation/evolution controversy was what encouraged Fix to examine more closely the claims of both sides of the debate. His research into these claims led him to reject assertions by both the scientific establishment and Christian fundamentalism. In other words, he rejects "the tired alternatives of Darwin and Genesis." He is particularly critical of paleoanthropology, which he considers "more of a market phenomenon than a disinterested scientific exercise." He is also disturbed by what he sees to be unwarranted claims about the fossil record as supportive of macroevolution. Fix rejects young earth creationism as the result of misinformed biblical literal-

ism but endorses the spiritual-agency-directed evolutionism promoted by Robert Broom. He, then, discusses "the realm of the spirit," delving into parapsychology, out-of-body experiences, and the like. He believes that forces he calls "psychogenesis" and "apparition theory" can account for the occurrence of all organisms.

Ford, Adam. *Universe: God, Science and the Human Person.* Mystic, CT: Twenty-Third Publications, 1987; 228 pp.
Ford, a priest and chaplain to a girls' school in London, evaluates important questions raised by the encounter between contemporary science and theology. Ford does not believe that science and theology are at war; instead, he thinks these two enterprises can be sources of enrichment for each other. He examines the relationship between modern cosmology and the creation story, the theological implications of evolution, the role of chance in creation, the nature of miracles, and other topics of interest to Christians interested in science. Ford is critical of traditional means of handling science/faith issues. He rejects biblical literalism, fundamentalist creationism, deism, reductionism, and "god-of-the-gaps" thinking. He believes that God's moment-by-moment activity upholds the natural world.

Frair, Wayne, and Percival Davis. *A Case for Creation.* Third edition. Chicago, IL: Moody, 1983; 155 pp.
Frair is a professor of biology at King's College, New York, and Davis teaches life science at Hillsborough Community College, Tampa, Florida. They wrote this book to show the informed layman "that evolutionary doctrine is wrong." The authors begin by discussing the nature of science and the history of scientific theories on origins. They believe that neither creationism nor evolutionism is scientific but that both positions are ultimately based in faith. They reject the notion that similar structures and enzyme homologies imply common descent. A discussion of earth history, the fossil record, and radiometric dating includes qualified support for flood geology, although the authors note that there are unsolved problems with this view. They question whether small-scale changes in organisms, such as the development of pesticide resistance, can really be considered to be examples of evolutionary change. They also examine human biology and the origin of behavior. In their last chapter, Frair and Davis discuss their biblical presuppositions: that God created the world in an orderly manner during a short period of time, and that humans

were the climax of this creation. The original created "kind," they think, may have been a much broader category than the modern "species." In an epilogue the authors express the need for more biblically based creationist scholars and researchers, people who are determined but also civil.

Frye, Roland Mushat (ed.). *Is God a Creationist? The Religious Case Against Creation-Science.* New York: Charles Scribner's Sons, 1983; 205 pp.
Many volumes have been written to counter the claims of creation science on scientific grounds. This book does so on the basis of biblical grounds. The editor is a professor of English at the University of Pennsylvania. Eleven other authors represent diverse academic fields and three faith traditions: Judaism, Protestantism, and Roman Catholicism. The first section of the book suggests that, in their zeal to defend God's creatorship, creation scientists misunderstand the relationship between science and the Bible and that any attempt to validate the biblical truth by means of science is misdirected. The second section argues that creation science is philosophically, theologically, and scientifically unsound, that flood geology is a fantasy, and that to substitute a scientific reading of the Genesis creation account for a symbolic meaning is a form of modernism. The third section affirms the Christian faith by considering the many evidences for design in the universe and by positing God's involvement in the evolutionary process itself. The fourth section includes articles by Jewish, Protestant, and Roman Catholic leaders, all of whom affirm their faith in God's creatorship. A concluding essay by Frye suggests that enlightened minds continue to be informed both by God's "Book of Nature" and his "Book of Scripture."

Geisler, Norman L. *Cosmos: Carl Sagan's Religion for the Scientific Mind.* Dallas, TX: Quest, 1983; 63 pp.
Geisler is a professor of systematic theology at Dallas Theological Seminary. This small book examines the nature of Carl Sagan's religious perspectives as promoted in his book, *Cosmos*, and in the television series by the same name. According to Geisler, Sagan is the most prominant spokesperson for "the religion of the COSMOS," primarily because he is such an effective and visible communicator. Geisler notes that Sagan's religion is atypical—it proposes no supernatural being, nor does it hold a set of writings as

authoritative. For Sagan, says Geisler, "the universe is a great cathedral, and scientific wonder and exploration are the attitudes of the devout." In contrast to Sagan, Geisler believes that the scientific evidence "pushes one back to the Creator of the cosmos."

_____. *Knowing the Truth About Creation: How It Happened and What It Means.* Ann Arbor, MI: Servant Books, 1989; 162 pp.
In this book, Geisler provides a description of the natural world from a nonscientific point of view. He also deals with more otherworldly topics, such as angels and heaven. In a discussion of how creation happened, Geisler distinguishes between what he calls "origins-science" and "operations-science." The former deals with "originating causality," whereas the latter is concerned with "conserving causality." He suggests that there are three models of origins: materialism, pantheism, and theism. Although he accepts Darwin's explanation for how the biological world works, he defends a theistic view of origins. Geisler believes that humans were given the ability to choose, and it is this capacity that gives them rulership over creation. This rulership comes with a responsibility to be proper stewards of the earth. He says that "if I am not the earth's keeper, then . . . neither am I my brother's keeper. For it is my brother's earth."

_____, and J. Kerby Anderson. *Origin Science: A Proposal for the Creation-Evolution Controversy.* Grand Rapids, MI: Baker, 1987; 198 pp.
In this book, the authors differentiate between empirical science and origin science. Empirical science "deals with observed regularities in the present" that are amenable to experimental verification. Origin science, by contrast, deals with one-time events that were not observed and cannot be repeated. The authors believe that the creation/evolution controversy "arises in part because of the confusion of [these two] different kinds of science." They suggest that "the big bang theory, which deals with a discontinuous singularity," opens the door "for a supernatural view" of the origin of the universe. The book was written for the layperson and can be understood without an extensive knowledge of science.

Giberson, Karl. *Worlds Apart: The Unholy War Between Religion and Science.* Kansas City, MO: Beacon Hill, 1993; 224 pp.
Giberson, a Nazarene physicist, sets out to examine the implications

of science for the Christian faith. He believes that problems between science and theology have arisen primarily as a result of a conflict of authority, not because the two are incompatible realms of thought. Beginning with the Greek philosophers, Giberson reviews the history of the science/religion conflict up until the present time. He is critical of creation science and believes that the best approach to resolving the conflict would be one that avoids scientific materialism, at one extreme, and biblical literalism at the other. He cautions readers not to jump to theological conclusions in the face of scientific pictures that are incomplete. He believes that, among other examples, the anthropic principle could provide a common ground of discussion for theologians and scientists. He respects evolutionary theory but also recognizes its many problems. Giberson writes autobiographically from his experience within the Nazarene community.

Gilkey, Langdon. *Nature, Reality, and the Sacred: The Nexus of Science and Religion.* Minneapolis, MN: Fortress, 1993; 204 pp.
Gilkey, a retired professor of theology at the University of Chicago, examines here the interrelation of science, theology, and nature. The thirteen chapters originally appeared as lectures and papers. Gilkey is concerned about maintaining a balance of perspectives in his approach to nature. He writes: "A purely 'religious' apprehension of nature, void of any influence of the scientific understanding of nature, is indefensible. . . . A purely 'scientific' apprehension of nature, void of any influence of the religious understanding of nature, is equally indefensible." In addition to a consideration of nature and its apprehension by modern science and religion, Gilkey contrasts contemporary scientific views of nature with those of "primal religions," believing that these archaic perspectives can "enrich and expand our own modern understanding."

Gilfillan, Marjorie Mary. *The Bible May Agree with Evolution.* Long Beach, CA: Wenzel, 1995; 306 pp.
Gilfillan is a researcher on folk dancing. She claims that dance costumes throughout the world share important similarities, and that these similarities provide evidence for the Genesis flood. She also posits, among other things, that evolution ended with the ice age, that the wives of Noah's sons were the orginators of the races, and that Adam was the hybridized offspring of Neanderthal and Cro-Magnon parents. This wildly speculative book is heavily footnoted.

Gish, Duane T. *Creation Scientists Answer Their Critics.* El Cajon, CA: Institute for Creation Research, 1993; 451 pp.

A large number of articles and books have appeared in recent decades seeking to discredit the creationist movement. Gish is one of the most outspoken and best known members of this movement. This book is his attempt to answer charges made by critics of creationism. He provides a brief historical overview and description of the terminology used in the creation/evolution controversy, followed by a consideration of the notion of scientific integrity. He examines the fossil record and thermodynamics, responding to criticisms against creationist claims in both of these areas. He takes particular aim at three anticreationist books: Philip Kitcher's *Abusing Science* (1982), Niles Eldredge's *The Monkey Business* (1982), and Laurie Godfrey's *Scientists Confront Creationism* (1983). A final chapter contains quotations used by creation scientists in their debates with evolutionists.

Ham, Kenneth A. *The Lie: Evolution.* El Cajon, CA: Creation-Life, 1987; 164 pp.

Ham is on the staff at the Institute for Creation Research. This book provides a series of arguments against the theory of evolution. He posits that both evolutionism and creationism are religious belief systems, and that people want to avoid accepting creationism, because by doing so they would be admitting that a creator makes the rules of life. He holds that sound Christian theology must be rooted in Genesis and that a belief in evolution over millions of years destroys the Christian doctrine of salvation through the death of Jesus. He believes that atheistic evolutionism is responsible for many of society's problems, including abortion, communism, Nazism, drug abuse, and homosexuality. He discusses the importance of the evolution/creation controversy in the last days of earth history and the importance of creationism to evangelism. He believes that creationists would be in a stronger position if they were to argue more from the point of view of presuppositions than from evidence. One appendix provides twenty reasons for why evolution and creation do not mix, while a second appendix addresses the question of why God took six days to create.

______, and Paul Taylor. *The Genesis Solution.* Grand Rapids, MI: Baker Books, 1988; 126 pp.

Ham and Taylor believe that the biblical stories of the creation, the fall, the tower of Babel event, and the flood are fundamental to all of Christian belief. They suggest that by eroding faith in the historicity of these events, Christians are destroying the foundation of their entire belief system. According to Ham and Taylor, evolutionism is responsible for the acceptance of homosexuality, abortion, premarital and extramarital sex, easy divorce, pornography, agnosticism, atheism, secular humanism, and other evils. The authors believe that creationist views like the day-age theory, progressive creationism, the gap theory, and other accommodationist interpretations of Genesis lead to the destruction of the Gospel message. Evolutionism, they say, is seeping into Christian churches, schools, and colleges. Christians, they suggest, should use "Creation Evangelism" to neutralize evolutionary philosophy before the gospel of Christ can be properly shared.

Hasker, William. *Metaphysics: Constructing a World View.* Downer's Grove, IL: InterVarsity, 1983; 132 pp.
Metaphysics is a branch of philosophy that addresses issues about the nature of reality—issues basic to those of the creation/evolution controversy. Hasker is a Christian professor of philosophy at Huntington College. In this book he describes the concerns of metaphysics. He posits that metaphysical questions are answered by providing good reasons for what we believe. He defends "emergentism," the view that the soul emerges as a result of the activities of the nervous system. He advocates what he calls a scientific picture of the world, which, although it may not provide a complete picture of reality, provides a picture that cannot be ignored. Hasker accepts theism, the view that God created the universe distinct from himself, and rejects naturalism, pantheism, and panentheism. He notes that any Christian metaphysic must assume that God is the ultimate reality, that God created reality outside himself, and that humans are made in God's image.

Haught, John F. *The Cosmic Adventure: Science, Religion, and the Quest for Purpose.* New York: Paulist Press, 1984; 184 pp.
Haught is an associate professor of theology at Georgetown University. In this book he asks if the religious intuition that the universe has purpose is consistent with the discoveries of contemporary science. He answers this question affirmatively. In fact, he believes that Christians can endorse and foster the scientific endeavor. He discusses a variety of interesting concepts, including the notion of chance and

its relationship to the development of novelty, the existence of beauty as an outgrowth of chaos and predictability, and the emergent qualities of higher organization levels. According to Haught, God is the highest emergent reality, and Jesus is the symbol through which the ultimate meaning of the universe becomes apparent to believers. Moreover, science is the means by which humans "grasp what lies below consciousness. . . . Religion, on the other hand, complements science by relating us to fields, dimensions or levels that lie above, or deeper than, consciousness in the cosmic hierarchy."

_____. *Science & Religion: From Conflict to Conversation.* Mahway, NJ: Paulist Press, 1995; 203 pp.
In this book, Haught introduces nonspecialists to basic issues in science and religion. Haught discusses five ways in which different people respond to conflicts between science and religion. The first response is *conflation*, the uncritical merging of religion with a few scientific ideas that are poorly understood. The other four ways are corrections to this first ill-informed response: with *conflict,* religion and science are seen as completely opposed to one another; with *contrast,* science and religion are seen as so different from one another, they could not possibly be in conflict; with *contact,* science and religion are viewed as distinct but as having implications for one another; and with *confirmation,* religion supports and blesses scientific endeavor. Haught is most comfortable with the last two of these responses. He believes that theology needs to be recast in evolutionary terms. He also suggests that "the prospect of mind's evolving" may have helped to shape the cosmos from the beginning.

Hayward, Alan. *God's Truth: A Scientist Shows Why It Makes Sense to Believe the Bible.* Nashville, TN: Thomas Nelson, 1983; 331 pp.
Hayward is a research and development advisor with Redwood International, Ltd. His intent in this book is to examine the claims of the Bible in terms that anyone can understand. Part I provides reasons for believing in the Bible as God's Word. These reasons include the fulfillment of prophecy, the life of Jesus, the timeless relevance of Old Testament law, and the internal harmony of scripture. Part II examines some of the claims of the Bible, particularly in light of science. Hayward believes that the "whole Bible stands or falls together"—that one cannot, for example, deny the historicity of Adam and accept the life of Jesus. On the other hand, he argues vigorously for accepting

the age of the earth as very old and for rejecting flood geology. Genesis 1, he believes, paints a broad outline of earth history. He thinks that the days of creation should be interpreted as the days on which God revealed creation. He vigorously opposes the theory of biological evolution. Part III encourages faith and provides suggestions for Bible study.

Henderson, Charles P., Jr. *God and Science: The Death and Rebirth of Theism.* Atlanta, GA: John Knox Press, 1986; 186 pp.
Henderson, a Presbyterian pastor and assistant dean of the Chapel at Princeton University, sets out to offer new and updated "proofs" for the existence of God. After evaluating the views of Albert Einstein, Sigmund Freud, Charles Darwin, Karl Marx, Teilhard de Chardin, Paul Tillich, and others, Henderson provides his own insights into issues of science and faith. Some of his assertions will appear odd to readers. For example, he suggests that before his death, the prominent Christian theologian Dietrich Bonhoeffer had become a scientific atheist, and he believed that "all forms of sexual expression are merely repressed spirituality." His ultimate purpose is to suggest that when "faith is internally consistent, coherent, and responsive to new insights which arise at the forward frontier of knowledge, then one has in fact established a new proof for God."

Hoover, A. J. *The Case for Teaching Creation.* Joplin, MO: College Press, 1981; 84 pp.
Hoover, a professor of history, begins this little book by defining what he means by "strict science" and "loose science." Strict science encompasses what can be observed and repeated through experimental verification. By contrast, loose science is based on inferences from circumstantial evidence. According to Hoover, theories of origins fit this latter category. In view of this, Hoover recommends that both creationism and evolutionism be presented to public school pupils. His third chapter defines what to him constitutes the creation model of origins—standard scientific creationism. He considers the creation stories of non-Christian cultures to be essentially evolutionary by nature and does not discuss the gap theory, progressive creationism, theistic evolution, and other biblically informed views. His final chapter recommends that "both creation and evolution should be investigated in classes like 'Theories of Origins,' just as we have classes called 'Theories of Personality' and 'Theories of History.' "

Huchingson, J. E. (ed.). *Religion and the Natural Sciences: The Range of Engagement.* Orlando, FL: Harcourt, Brace, Jovanovich College Publishers, 1993; 402 pp.

This anthology provides a wide-ranging view of perspectives on issues in science and religion by many prominent authors including Martin Buber, Pierre Teilhard de Chardin, Albert Einstein, Duane Gish, C. S. Lewis, John Polkinghorne, and Edward O. Wilson. The 48 readings are relatively short. Topics include the points of engagement between science and religion, types of language used by people of science and people of faith, the cosmos and design, creation and evolution, sociobiology, and the impact of faith on views of the natural world. This volume provides an excellent sampling of the views on issues related to science and religion; readers will need to look elsewhere, however, for detailed expositions of these perspectives. Each chapter ends with a series of questions for study, questions for reflection, and suggestions for networking among the various papers.

Hummel, Charles E. *The Galileo Connection: Resolving Conflicts Between Science and the Bible.* Downers Grove, IL: InterVarsity Press, 1986; 263 pp.

Hummel, who has graduate degrees in both chemical engineering and biblical literature, wrote this book with Christian college students in mind. His goal is to show that there is no real conflict between Christianity and science. Hummel first provides carefully researched discussions of the lives of four important scientists who were Christians: Copernicus, Kepler, Galileo, and Newton. He then discusses basic principles of biblical interpretation, concluding that it is a mistake to try to derive scientific information from scripture. Instead, he believes that science and the Bible provide complementary perspectives on reality, neither of which is complete without the other. Hummel is critical of the so-called "concordist" approach to Genesis 1, which attempts to harmonize the creation story with modern science. He believes that Genesis 1 should be interpreted as it was originally intended: as a polemic against polytheism and the deification of nature.

Huse, Scott M. *The Collapse of Evolution.* Grand Rapids, MI: Baker, 1983; 178 pp.

Huse wrote this book to highlight scientific problems with evolutionary theory, offer scientific support for Biblical creationism, and demonstrate that evolution and creationism are mutually exclusive concepts. Huse believes that the theory of evolution is a scientific blunder and that theistic evolution is an indefensible compromise position between evolutionism and creationism. He posits that acceptance of both evolution and creation is ultimately based on faith but that scientific evidence regarding the past should not be overlooked. He reviews evidence from anthropology, biology, geology, and other sciences cited by Henry M. Morris in support of creationism. He also critiques the use of *Archaeopteryx,* fossil horses and the results of origin of life experiments to support evolutionary theory. Huse lists the names of creationist scientists and in an appendix provides a list of creationist organizations.

Hyers, Conrad. *The Meaning of Creation: Genesis and Modern Science.* Atlanta, GA: John Knox, 1984; 203 pp.
Hyers is a professor of religion at Gustavus Adolphus College in Minnesota. This book addresses the meaning of the biblical creation stories in the context of the creation/evolution controversy. Specifically, Hyers is concerned about the most appropriate way to read the Genesis text. He believes that any attempt to evaluate Genesis in the context of modern science, either positively or negatively, is misguided effort. The Genesis text must, instead, be interpreted on its own terms and in the context of ancient Middle Eastern culture. Hyers views the Genesis accounts as mythological and symbolic in the highest sense of the word. He believes that use of such terms is "essential in avoiding literal reductionism and a trivialization of the Genesis materials." A literalistic interpretation of the Genesis text, he believes, robs it of its power, nuance, and mystery; moreover, with literalist approaches "its mood of celebration and its significance are largely lost in the clouds of geological and paleontological dust stirred up in the confusion."

Jaki, Stanley L. *Angels, Apes, & Men.* LaSalle, IL: Sherwood Sugden, 1983; 128 pp.
Jaki, a Benedictine priest, holds doctorates in both theology and physics. This book is a collection of his lectures given at the Institute for Christian Studies, Toronto. Jaki is concerned that scientists avoid an overdependence on rationalism, on one hand, and empiricism on the other. In these lectures he is concerned with how scientists and

philosophers approach the nature of man. According to Jaki, people like Descartes, Kant, and Hegel were "Angels" who emphasized rationalist views of man, views that did not take into account the multiple dimensions of human nature. By contrast, thinkers like Rousseau and Darwin were "Apes" who emphasized an empiricist approach to humans, seeing them as mere animals. According to Jaki, both extremes mistake the true nature of humans. In his view, humans indivisibly consist of both body and mind.

Johnson, Phillip E. *Reason in the Balance: The Case Against Naturalism in Science, Law and Education.* Downers Grove, IL: InterVarsity, 1995; 245 pp.
Here attorney Johnson, whose earlier book *Darwin on Trial* (1991) provided a scathing critique of Darwinism, develops a similar and related case against naturalism and its domination over contemporary science, law, politics, and society. He begins by asking if God is "unconstitutional," suggesting that recent legal decisions and pronouncements by the intellectual elite would lead one to think so. He argues that "metaphysical naturalism" has become the established "religious" philosophy of American academics, a philosophy that exerts a powerful monopoly over competing systems of thought. Metaphysical naturalism, he believes, is rooted in the triumph of Darwinism in science but influences every other aspect of human society as well. Johnson takes the position that naturalism should be "tried" in the intellectual courtroom and leaves open the possibility that metaphysical naturalism might be right—that ultimately the universe has no purpose. He clearly believes, however, that this would not be the outcome. He is particularly critical of theists who subscribe to "methodological naturalism"—that in the actual practice of science, one cannot assume the possibility of God's involvement in nature. He argues, instead, for "theistic realism" which assumes that the consequences of God's activites are detectable by the methods of science, assuming the right questions are asked.

Kaiser, Christopher. *Creation and the History of Science.* Grand Rapids, MI: Eerdmans, 1991; 316 pp.
Kaiser, educated in both theology and astrophysics, believes that contemporary scientific creationists have strayed from historic interpretations of the Christian doctrine of creation. Kaiser outlines four positions that he believes constitute the core of the historic creation doc-

trine: (1) the universe is comprehensible because its creator also designed human reasoning capabilities; (2) the heavens and the earth are part of the same creation and operate according to the same principles; (3) nature possesses "relative autonomy," meaning that it operates on the basis of created laws; and (4) humans should use their understanding of the world to benefit others. Kaiser reviews the development of these themes throughout Christian history and explains how their development depended on an interaction between theology and science. Ultimately, Kaiser holds that to believe in creation means to believe that the universe exists because of God's will, action, and establishment of natural law at the beginning of time.

Kealy, Seán P. *Science and the Bible.* Dublin: Columba Press, 1987; 91 pp.

In this small book, Kealy provides a collection of nontechnical essays on the relationship between science and the Bible. He deals with issues such as creation, evolution, miracles, and shifts in human understandings of the universe. Kealy believes that the Bible is God's inspired word, but also accepts the standard interpretations of science, including geological and biological evolution, as our best approximations of reality. He believes that today's scientists are becoming more and more open to the notions of mystery and transcendence, and that scientists and theologians are developing better mutual respect thereby increasing the effectiveness of their communication.

King, Thomas M., and James F. Salmon (eds.). *Teilhard and the Unity of Knowledge.* Ramsey, NJ: Paulist Press, 1983; 172 pp.

This collection of essays represents the proceedings of the Georgetown University Centennial Symposium honoring French paleontologist and theologian Teilhard de Chardin. Participants include such notables as paleoanthropologist Richard Leakey and economist Kenneth Boulding. They write on a loose variety of themes related to human evolution, including the non-equilibrium system development, human time-consciousness, esthetics, and evolution in relation to science and faith. An appendix details Teilhard's participation in the Piltdown hoax.

Klotz, John W. *Studies in Creation: A General Introduction to the Creation/Evolution Debate.* St. Louis, MO: Concordia, 1985; 216 pp.

Klotz is a professor of natural science at Concordia College, Indiana. This book contains many of the same themes presented in his earlier book, *Genes, Genesis and Evolution* (1955). He first examines the nature of science, scientific methodology, and the theory of evolution. He then discusses creation from a biblical point of view and reviews various approaches to the interpretation of Genesis. After providing a general overview of evolutionary theory, Klotz devotes his next-to-the-last chapter to problems confronting creationists, such as evidence from the fossil record, the data on plant and animal distribution patterns, and the theory of continental drift. His last chapter is devoted to difficulties for evolutionists—the lack of an adequate mechanism for the spontaneous origin of life, the incomplete fossil record for human evolution, the problem of speech development, fossil frauds, and the generally harmful effects of mutations, among others. Klotz believes that "the creationist is free to postulate either a young or an old earth." Indeed, throughout the book he assumes a relatively non-dogmatic and informed, though biblically conservative, approach to origins.

Kraft, R. Wayne. *A Reason to Hope: A Synthesis of Teilhard de Chardin's Vision and Systems Thinking*. Seaside, CA: Intersystems, 1983; 274 pp.
Kraft is a professor of metallurgy and materials engineering at Lehigh University. An admirer of Teilhard's views on science and Christianity, Kraft seeks to bring these two themes together within the context of the French paleontologist's system of thought. Kraft first addresses the issue of time, noting that it must progress given the process of evolution and the second law of thermodynamics. Broaching the issue of evolution, he posits that no "educated and rational person" today can deny that humanity "is the product of some sort of evolution." He also examines the concepts of entropy, systems thinking, communication, and energy. He finishes the book with a chapter called "Christogenesis," the last stage of Teilhard's process of making the world into God's image through the transforming power of God's love. Although Kraft is completely committed to Christianity, he believes that other world religions contain important truths that can enrich Christianity.

Levenson, Jon D. *Creation and the Persistence of Evil: The Jewish Drama of Divine Omnipotence*. San Francisco, CA: Harper & Row,

1988; 182 pp.
Levenson, who teaches Hebrew Bible at the University of Chicago, addresses three problems with standard interpretations of the Genesis creation text. First, he believes that the concept of creation out of nothing—*creatio ex nihilo*—is "not an adequate characterization of creation in the Hebrew Bible." Second, he contends that the first version of the creation story in Genesis 1:1-2:3 has not been adequately understood in relation to the priestly theological tradition, often thought to be the source of this account. Finally, he posits that the relationship between the idea that God is Creator and the notion that God is Lord needs further exploration. He addresses these problems in light of other Near Eastern creation texts and renditions of the creation story in Rabbinic Judaism. In short, he attempts to lead readers into a better understanding of the original meaning of the Old Testament creation texts.

Lucas, Ernest. *Genesis Today: Genesis and the Questions of Science.* London: Christian Impact, 1995; 160 pp.
Lucas, a former research fellow in chemistry at Oxford University, teaches at Bristol Baptist College. He wrote this book to provide a proper understanding of Genesis 1-11 in relation to modern science. He notes that both science and the Bible are frequently misunderstood. Scientists have sometimes made the mistake of blurring fact and theory, though Lucas posits that they are becoming more aware of this problem. He also notes that science cannot be carried out in a purely objective vacuum—every scientist has faith, even if it is only in his or her rationality and in rationality of nature. In addition, scientists are selective in their collection of data. Likewise, religious people often misunderstand the source of their "data," the Bible. Lucas believes that each part of the Bible should be read with an eye toward the literary character of the text and that wholesale literalism should be avoided. He believes that the Genesis creation story was written as a polemic against paganism, not as a historical account. He argues that religious faith need not be irrational and that a positive interaction can occur between science and faith.

Maatman, Russell. *The Impact of Evolutionary Theory: A Christian View.* Sioux Center, IA: Dordt College Press, 1993; 318 pp.
Maatman, a retired professor of chemistry at Dordt College, has a long-standing interest in the relation between science and the Chris-

tian faith. In this book, he examines the history of evolutionary theory and its interactions with Christian theology. He discusses the design argument and various objections to it, both theological and scientific. He then provides an overview of modern evolutionary theory and surveys some of the problems with this theory. After discussing the concept of revelation and how it relates to science, Maatman examines seven different interpretations of Genesis 1 and 2. The final three chapters evaluate the impact of evolutionary theory on diverse areas of human experience, from history and religion to economics, feminism, and animal rights. This book represents an interesting perspective from a scientifically informed Christian who believes in a real Adam and Eve, but who also believes that the earth and life are millions of years old.

MacKay, Donald. *Science and the Quest for Meaning*. Grand Rapids, MI: Eerdmans, 1982; 75 pp.
The contents of this little book were first presented as the Pascal Lectures at the University of Waterloo. Specifically, MacKay addresses the issue of Christian meaning in light of the scientific process. He does not believe that science destroys meaning but rather that it enhances meaning. Also, he does not believe that the notions of chance and reductionism, championed by science, destroy the credibility of the Christian faith. Indeed, he posits that the Christian view of an ordered universe is what gives science its rationale. He suggests that increased scientific knowledge leads to increased awe, as well as to increased accountability for what we know. Discussions that followed MacKay's lectures are also provided.

Mangum, John M. (ed.). *The New Faith-Science Debate: Probing Cosmology, Technology, and Theology*. Minneapolis, MN: Fortess Press, 1989; 165 pp.
This volume contains the proceedings of the 1987 Cyprus consultation sponsored by the Lutheran Church in America and the Lutheran World Federation. Papers deal with topics such as the nature of modern science, the integration of modern science and theology from a process theology perspective, the church's responsibility to teach people science and technology as a part of their liberation, the impact of "high-tech" on American society, the challenges of genetic engineering and its role of "co-creation," the relationship of Asian religions and science, the practice of science as a Christian vocation, and

the importance of a four-way conversation among the disciplines of science, theology, technology, and ethics.

Margenau, Henry, and Roy Abraham Varghese. *Cosmos, Bios, Theos: Scientists Reflect on Science, God, and the Origins of the Universe, Life and Homo Sapiens.* LaSalle, IL: Open Court, 1992; 285 pp.
This book presents essays by some sixty scientists, including twenty-four Nobel laureates, who respond to specific questions about the origin of the universe, life, and humans in view of religious faith. Many of the authors say they believe in the existence of a creator because the universe seems to be put together so well.

Mitchell, R. G. *Einstein and Christ: A New Approach to the Defense of the Christian Religion.* Edinburgh: Scottish Academic Press, 1987; 231 pp.
Mitchell offers an apology for the Christian faith from a Catholic Thomistic perspective in relation to modern science. He writes for non-Christians and for students who are questioning their faith. Mitchell believes that a relevant Christian faith is one that takes into account contemporary knowledge. He believes that humans were created through a directed evolutionary process and that the cosmos was designed to provide a home for humankind. He finds parallels between the teachings of Christ and Einstein's theory of relativity. He discusses the works of other prominent scientists and philosophers and finds support for a Christian perspective on the cosmos. Mitchell hopes that scientists and theologians can become open-minded enough to engage in productive dialogue with each another.

Montgomery, Ray N. *It's a Wonderful World—Naturally.* Washington, DC: Review and Herald, 1982; 128 pp.
Montgomery is a retired teacher with special interests in biology, library science, and religion. This book provides a collection of facts and anecdotes from nature that "is surely evidence of a remarkable Creator who possesses a sense of the humorous, as well as the awe-inspiring and delightful." He discusses the lives of elephants, shrews, primates, dogs, trees, flowers and other organisms as evidence for divine design. He notes that "Some Christian naturalists believe that a few trees now living began to grow at Creation, and in a sense have botanical immortality." While he says that animals have no way to keep the first three of the Ten Commandments, he suggests that they

operate in accord with the last seven. Bad weather and disastrous storms, he believes, began after the flood; moreover, half the water of the flood came from vast caverns, the "fountains of the deep," beneath the surface of the earth. In fact, many creationists, he says, "believe that the spectacular caverns, such as Luray, Mammoth Cave, and Carlsbad Caverns, are entrances to the original fountains of the great deep." Montgomery writes that "in the earth made new," after the return of Christ, lions, tigers, and bears will be herbivorous and without canines and "upper teeth in the front of the mouth."

Moreland, J. P. *Christianity and the Nature of Science: A Philosophical Investigation.* Grand Rapids, MI: Baker, 1989; 263 pp.
Moreland, a professor of philosophy at Liberty University, evaluates the nature of science with a view toward demonstrating that creation science is truly science. Following a discussion of definitions of science, Moreland suggests that science has no particular right to exclusive knowledge over other disciplines because of its particular methodology. This is followed by a discussion of the limitations of science and an assessment of the view that scientific theories are real or approximately real models of the universe. He suggests that when science and theology come into conflict, resolution may be achieved by assuming an antirealist view of the science. Finally, Moreland discusses the scientific status of creationism. He believes that the creation/evolution debate is largly "a conflict over epistemic values."

Morowitz, Harold J. *Cosmic Joy and Local Pain: Musings of a Mystic Scientist.* New York: Charles Scribner's Sons, 1987; 303 pp.
Morowitz, a prominent biophysicist from Yale University, attempts in this book to "bridge the gulf between religion and science." He writes from a sailboat off the coast of Hawaii where he has time to ponder the deep questions of meaning. Morowitz, a man of Jewish heritage, describes himself as a pantheist—he finds god in nature. He discusses the remarkable order, interconnectedness, and apparent design in nature which form the basis of his religion. According to Morowitz, the practice of science is the means by which we become "partners with god in making the future." Morowitz writes in a nontechnical style for the lay reader.

Morris, Henry M. *The Biblical Basis for Modern Science.* Grand Rapids, MI: Baker, 1984; 516 pp.

In this lengthy book Morris, founder and president of the Institute for Creation Research, uses the Bible as a basis for interpreting a broad range of scientific disciplines. For example, he discusses Biblical astronomy, Biblical thermodynamics, Biblical chemistry, Biblical geophysics, Biblical paleontology, Biblical biology, and Biblical ethnology. Also included are chapters on theology, the "Queen of Sciences," and evolutionism, "Science Falsely So Called." Included are many of the standard creation science topics: thermodynamics, flood geology, canopy theory, Noah's ark, and so on. Morris assumes a literalistic interpretation of the Bible throughout. He posits that the universe consists of three parts—space, time, and matter—an analogy for the three-part Godhead—Father, Son, and Holy Spirit.

_____. *Creation and the Modern Christian.* El Cajon, CA: Master Books, 1985; 266 pp.
Morris directs this book to fellow Christians, urging them to do battle against "evolutionary humanism" and encouraging them toward a creationist revival. He sees virtually every modern evil—including racism, imperialism, and communism—as an outgrowth of evolutionary humanism, a force seeking to control modern society. By contrast, he believes that creationism is the source of Americanism. He is critical of the principle of uniformity, historical geology, thrust fault theory, geologic dating techniques, and other aspects of earth science. This is a call to war by America's most prominent creation scientist spokesman.

_____. *Science and the Bible.* Chicago, IL: Moody Press, 1986; 154 pp.
This volume is an updated version of the author's *That You May Believe* (1946), later published as *The Bible and Modern Science* (1951). As an outgrowth of Morris' evangelistic fervor, he hopes this book will "win people to a genuine faith in Jesus Christ." Morris begins with an examination of what he calls the Bible's "scientific accuracy." He discusses a variety of biblical texts that he believes are confirmed by modern astronomy, geophysics, hydrology, and biology. He also considers principles of matter and energy, as well as the question of miracles. Morris then addresses the theory of evolution, which he believes to be utterly false and indefensible. He opts, instead, for a belief in a six day creation that occurred less than ten thousand years ago. He then broaches the topic of earth history and

asserts that flood geology provides a better explanation for geologial evidence than standard geological theories. His final two chapters discuss how archaeology, ancient history, and fulfilled prophecy affirm the trustworthiness of scripture.

_____. *Remarkable Record of Job.* Santee, CA: Master Books, 1988; 146 pp.
One reason Morris values the biblical book of Job is that he believes it supports a literal interpretation of Genesis 1-11 and that it even adds details to the story of creation. He also sees Job as providing insights into Satan and his activities. In addition to comments on the words of Job's friends, Morris delights in highlighting allusions to creation and earth history in Job. He notes that there are more references to the flood in Job than references to creation. This, he says, is because "Job's experience could have occurred only 300 or so years after the flood." He interprets mention of the "behemoth" and the "leviathan" in Job as references to now extinct animals, possibly dinosaurs, and passages referring to "cold," "snow," and "ice" as "hints of the post-flood Ice Age."

_____. *The Long War Against God: The History and Impact of the Creation/Evolution Conflict.* Grand Rapids, MI: Baker, 1989; 344 pp.
Morris contends that the evolution/creation controversy has its roots in the original rebellion of Satan against God. Morris sees modern evolutionism as simply a continuation of that rebellion. He believes that evolutionism is virtually synonymous with atheism, naturalism, humanism, and materialism, and that it has been responsible for Marxism, Nazism, communism, racism, abortion, a general decline in morality, and everything else that is evil in the world. Morris is amazed that a book such as Darwin's *Origin of Species,* "which "is most notable for its complete lack of documentation," has had such a "profound influence of the subsequent history of human life and thought." According to Morris, the acceptance of evolution is not made on the basis of scientific evidence but on the basis of one's choice of a worldview. Special creation, says Morris, is the most certain truth of science.

_____. *Biblical Creationism.* Grand Rapids, MI: Baker, 1993; 276 pp.
In this book, Morris examines explicit and implicit references to

creation and the flood in each book of the Bible, as well as in the Apocrypha, the Pseudepigrapha, and the works of Josephus. While he believes that the extra-biblical sources include some untrustworthy statements about biblical history, they nonetheless present an uncompromised, truthful view of creation and the deluge. Morris argues that the many allusions to creation and the flood throughout scripture make it clear that the Genesis accounts of these events should be taken literally. Moreover, Christians who seek to accommodate the biblical position to modern scientific theories are simply trying to appease their non-Christian colleagues. He posits several untestable creationist hypotheses, including that Adam spoke Hebrew, that the stars serve to house angels, and that the law of gravitation began to operate on the second day of creation. Morris believes that effective Christian evangelism must emphasize creationism.

_____, and Gary E. Parker. *What Is Creation Science?* San Diego, CA: Creation Life, 1982; 306 pp.
In this book, Morris and Parker describe the belief structure of creation scientists. They suggest that creationism is not only central to "conservative Protestantism, but also traditional Catholicism and Orthodox Judaism, as well as conservative Islam and other monotheistic religions." Creationism, they write, posits that the universe was created by processes not occurring today. They also hold that a creator would have been necessary for production of the first DNA and protein molecules; that homology and common embryological pathways are best viewed as evidence for common design, not common ancestry; that only variation within type ("subspeciation") has been observed, not change from one type to another ("transpeciation"); that fossil human footprints have been found with those of dinosaurs; that the Grand Canyon is not a slice through time but a transect through ancient ocean zones; that thermodynamic principles render evolution impossible; that the fossil record contains no transitional series; that the geologic column is found only in textbooks, not in nature; and that organisms represented as fossils lived "concurrently in one year." Appendices answering criticisms leveled against creation science conclude this book.

Morton, Glenn R. *Foundation, Fall and Flood: A Harmonization of Genesis and Science.* Dallas, TX: DMD Publishing, 1995; 159 pp.
Morton, a Christian geologist, believes that the Bible should be in-

terpreted literally, unless overwhelming evidence contradicts such an interpretation. A one-time young earth creationist, Morton was forced by the geological evidence to reevaluate his understanding of earth history. For him this meant acceptance of the evolutionary process over long ages. He argues that during each of the six, 24-hour days of creation God announced his intent of what he would create, then initiated the necessary processes for a gradual unfolding of the natural world over millions of years. All along the way, God remained involved in the creative process. According to Morton, humans originated around 5.5 million years ago by divine intervention. When Genesis states that a particular patriarch became the father of another patriarch, this simply means the first individual was an ancestor of the second individual mentioned. Morton sees no evidence for a universal flood between four and five thousand years ago. Instead, he posits that the flood story had its roots in the filling of the Mediterranean Sea some 5.5 million years ago.

Mott, Nevill (ed.). *Can Scientists Believe? Some Examples of the Attitude of Scientists to Religion.* London: James, 1991; 182 pp.
This book contains fourteen articles by religious believers of a variety of faiths, and one article by a nonbeliever. Positions espoused range from the view that religion is insignificant to science, to the notion that religious faith should have priority over science. Editor Mott denies most elements of the Christian faith, including miracles, the virgin birth, the resurrection, and an omnipotent God. He does, however, believe that the concept of God gives meaning to the existence of human consciousness, something that Mott sees as a great mystery. Other contributors deal with topics such as randomness, quantum mechanics, neuronal function, and design in nature in relation to theism. This book provides an interesting collection of widely divergent perspectives on faith in relation to views of the cosmos.

Murphy, George L., LaVonne Althouse, and Russell Willis. *Cosmic Witness: Commentaries on Science/Technology Themes.* Lima, OH: CSS Publishing, 1996; 183 pp.
This book is a contribution by the Evangelical Lutheran Church in America's Work Group on Science and Technology, the Working Group on Science and Technology of the United Church of Christ's Board of Homeland Ministries, and the Ecumenical Roundtable on

Science and Technology. It is designed to help preachers find scriptural "resources to address scientific and technological concerns," including those related to environmental quality and the history of life. The authors believe that even if the universe and its history can be completely explained from a naturalistic perspective, it "remains a work of divine creation." Accordingly, scientific explanation neither eliminates nor requires God's involvement. In accordance with the views of Teilhard de Chardin, the authors are not opposed to seeing the evolutionary process as ordained by God to lead to Christ. They believe that the "Technology of redemption and renewal is technology which promotes justice, sustainability, and participation." Appendices include indexes of biblical texts; sermon topics dealing with science, technology, and justice; and two sermons provided as examples of imaginative preaching.

Nee, Watchman. *The Mystery of Creation.* New York: Christian Fellowship, 1981; 149 pp.
Nee is a Chinese Christian who has written over thirty devotional books. This volume contains essays originally published in *Christian Magazine* from 1925 to 1927. Nee believes that every word of the Bible is "God-breathed," that "God knows everything," and that geology "is man's invention." Therefore "If Genesis and geology differ," writes Nee, "the error must be on the side of geology, for the authority of the Bible is beyond questioning." He believes that Genesis 1:1 refers to an initial perfect creation containing pre-Adamic humans; that because of Satan's fall, this initial creation was laid waste; and that during this time, the fossil record and geologic column were formed. According to Nee, God then recreated the word in six literal days around six thousand years ago; Adam and Eve were part of this recreation. No species of plant or animal, he believes, can become another species of plant or animal. Nee views the physical creation as a metaphor of the spiritual creation, and the last half of the book draws numerous parallels between the two.

Paine, S. Hugh. *Founded on the Floods.* Walnut, CA: Productions Plus, 1993; 150 pp.
Paine, a former metallurgist at Argonne National Laboratory and a retired professor of physics from Houghton College, wrote this book to document his insights into the Genesis record of earth history. Paine spent a number of years studying ancient Hebrew so he could read the

original Genesis text himself. As a result he came to the conclusion that one must read the text in its original language to gain a true appreciation of its meaning. While the Bible sometimes uses figurative language, he believes that it generally means what it says and must be taken as authoritative history. He also believes that verified scientific data must also be taken at face value. Paine believes that Noah's flood was universal but placid, and that earth history can best be understood in terms of the Gap theory and the flood story. He discusses the age of the earth and universe, the geologic column, and pre-Adamic hominids in relation to his views.

Pannenberg, Wolfhard. *Toward a Theology of Nature: Essays on Science and Faith,* Ted Peters, editor. Louisville, KY: Westminster/John Knox, 1993; 166 pp.
Pannenberg is a well-known professor of systematic theology at the University of Munich. This book is an attempt to address some of the philosophical issues surrounding the science/faith discussion. Specifically, Pannenberg encourages scientists to allow for the presence of God in their models of nature, while at the same time encouraging theologians to take the findings of science into consideration as they formulate their views. He begins with a series of theological questions for scientists. This is followed by discourses on the doctrine of creation in light of modern science, the interrelation of God and nature, contingency and natural law, the task of theology in relation to the divine spirit and nature, and the relation of spirit to energy and mind. As examples of his attempt to bridge theology and science, he suggests that the Christian doctrine of the Trinity could be invigorated by connecting the concept of Logos with information theory and by recognizing the activity of the divine spirit in the process of evolution. This is challenging and deeply philosophical work by a noted scholar.

Paul, Iain. *Science and Theology in Einstein's Perspective.* Edinburgh: Scottish Academic Press, 1986; 109 pp.
"Modern science and systematic theology have a common concern with all that lies between the birth of the universe and the end of the world," notes the author, who earned doctorates in both physical science and systematic theology. He attempts to show that modern science and theology share many similarities in terms of faith, knowledge, communication, authority, and rationality. This book is de-

signed to provide a framework in which theologians and scientists can communicate intelligently with one another.

Peacocke, Arthur. *Intimations of Reality: Critical Realism in Science and Religion.* Notre Dame, IN: Notre Dame Press, 1984; 94 pp.
Peacocke is a biochemist and dean of Clare College, Cambridge University. This book consists of two Mendenhall Lectures delivered at DePauw University in 1983. In his first lecture, "Ways to the Real World," Peacocke examines various philosophical approaches to scientific knowledge, including naive realism, logical positivism, and critical realism. He favors the last view, critical realism, which suggests that if a scientific theory is successful over the long term, then the theory may represent some aspects of reality. He discusses the use of models and metaphors in science and theology and notes that the very use of these devices implies that we are not providing a literal description of reality. In his second lecture, "God's Action in the Real World," Peacocke posits that theological constructs must be formulated with the best of our scientific knowledge in mind. He believes that science and theology can and should interact with one another as both seek truth. He accepts the notion of "panentheism," that "God includes and penetrates the whole universe, so that every part of it exists in him, but that his Being is more than, and is not exhausted by, the universe." Peacocke attempts to meld his theological insights with the theory of biological evolution and the operation of chance as a creative agent in the universe.

_____. *Theology for a Scientific Age: Being and Becoming—Natural, Divine, and Human.* Minneapolis, MN: Fortress Press, 1993; 416 pp.
In this book Peacocke attempts to create a theology responsive to a scientific understanding of the world. In Part I, he addresses several important concepts, including irreducible levels of organization in nature, "top-down" causation, and human personhood. Part II discusses the author's views on the nature of God's interaction with the world in view of the concepts addressed in Part I. Part III, an outgrowth of the author's Gifford lectures at Saint Andrews University in 1993, is concerned with the nature of humanity and God's revelation to humans. Peacocke believes that God works through cause and effect to bring about his will for the world. Consequently, Peacocke is critical of interventionist views of God's relationship with the world.

_____ (ed.). *The Sciences and Theology in the Twentieth Century.* South Bend, IN: Notre Dame Press, 1986; 309 pp.
The fourteen papers composing this volume were first presented in 1979 at an international symposium at Oxford, England. The authors examine to what extent "the common intention to seek intelligibility in human life and its surroundings" results in the mutual modification of different human enterprises, such as science, theology, and philosophy. The book is laden with philosophical and theological jargon making it accessible primarily to specialists interested in this topic.

Pearcey, Nancy R., and Charles B. Thaxton. *The Soul of Science: Christian Faith and Natural Philosophy.* Wheaton, IL: Crossway, 1994; 298 pp.
Pearcy is a science writer and Thaxton is a chemist. Pearcy and Thaxton argue that science is not at cross purposes with religion, but that science is actually rooted in religious and philosophical thought. The universe, they say, was created by God in an orderly way; therefore scientists should be able to evaluate the universe through the orderly scientific method. Various approaches to reality are discussed, including reductionism, pragmatic realism, and subjective relativism. Pearcy and Thaxton posit that people interpret scientific data in the context of three different theoretical frameworks: Aristotelian, neo-Platonic, and mechanistic. Historical examples of scientists representing each of these three perspectives are presented. The authors contend that a scientist's metaphysical assumptions actually guide her/his research, including the methodology used and the types of data gathered. The nature and implications of mathematics, relativity, and quantum mechanics are discussed in relation to worldviews. A final chapter examines DNA as an information molecule, as well as the irreducible structure of life. Extensive notes and suggested reading are provided.

Pilkey, John. *Origin of the Nations.* San Diego, CA: Master Books, 1984; 346 pp.
Pilkey, a young earth creationist, earned advanced degrees in literature and theology. According to Pilkey, the "Bible functions as a book of salvation precisely because it is a book of science, literature, and philosophy." The pivotal event of Pilkey's history is Noah's flood, which he dates at 2150 B.C. He believes that "Noachic science" is the

"atomic physics of world history" and will eventually liberate humankind from Darwinism. Pilkey finds the roots of the "antichrist" in a rebellion against Noah. He believes that human racial diversity arose from Noah's wife and the wives of Noah's sons. While he suggests that humankind's apocalyptic sensitivities have been eroded by Darwinism, Pilkey posits that the apocalyptic tradition has been preserved by the "fictional and dramatic subgenres of historical romance, gothic fiction, murder mystery, and local color fiction." Pilkey's discussion of history and meaning in the context of biblical truth and "Noachic science" ranges broadly, with discussions of mythology, literature, philology, historiography, art, and philosophy.

Polkinghorne, John. *Science and Providence: God's Interaction with the World.* Boston, MA: Shambhala, 1989; 114 pp.
Polkinghorne is president of Queen's College, Cambridge, an Anglican priest, a Fellow of the Royal Society, and past professor of mathematical physics at Cambridge. He is concerned with the issue of theism in an increasingly secular, technological society. Polkinghorne raises the question: "What would constitute evidence of God's activity?" If God is consistent and the basis of all that is, then it would be impossible to predict what it would be like without God. Polkinghorne believes that indeed God is the basis for all existence, but he also believes that "without some recourse to the particular there is a danger that the God who does everything will be perceived as the God who does nothing." He then examines the notion of God's activity in the world, as well as the concepts of providence, miracle, evil, prayer, time, incarnation, and hope. He patiently reasons his way through each topic. This book provides a thoughtful introduction to the view that God exists and interacts with the natural world.

_____. *Reason and Reality: The Relationship Between Science and Theology.* Philadelphia, PA: Trinity Press International, 1991; 104 pp.
This little book is a collection of eight essays, six of which are based on invited lectures by Polkinghorne. The author begins by looking at the relationship between science and theology, by establishing the basis for his "critical realism" approach to the subject, and by examining the use of metaphor in science and theology. He then discusses the nature of physical reality, which he views as a series of emergent levels of organization. Polkinghorne suggests that revela-

tion consists of an encounter with the Divine, not factual propositions of unalterable truth. He rejects both creationism, which imposes theological constraints on science, and scientism, which imposes scientific constraints on theology. He believes that theology can provide answers to "meta-questions" raised by science but which cannot be answered by science. In his final chapter on the Fall, Polkinghorne distinguishes "natural evil" from "moral evil" and suggests that while the entire universe has fallen physically, only part of it has fallen morally.

_____. *The Way the World Is.* London: Triangle, 1992; 130 pp.
Polkinghorne wrote this book soon after the start of his ministry to explain his religious beliefs to his "physics friends." In part, he attempts to show similarities between the methodology of natural scientists and theologians. This book is primarily a rational defense of Polkinghorne's belief in the New Testament gospel. While it provides an interesting look at how a prominent scientist views his faith, more complete discussions of science and religion can be found in his other books.

_____. *The Faith of a Physicist: Reflections of a Bottom-Up Thinker.* Princeton, NJ: Princeton University Press, 1994; 211 pp.
This book consists of a systematic theological treatment of these issues based on the Nicene Creed. Polkinghorne is disturbed by the tendency of many theologians to think deductively—from the "top-down." Instead, he is interested in the evidence that makes something seem true—a "bottom-up" or inductive approach. He is critical of process theology, pantheism, deism, and generalist notions of God as some type of Cosmic mind. He continues to take the Gospels, including the historicity of Jesus' virgin birth and resurrection, seriously. He believes that God is continually involved in upholding the creation, and that occasionally this involvement can be "exercised in specific ways."

_____. *Serious Talk: Science and Religion in Dialogue.* Valley Forge, PN: Trinity Press International, 1995; 117 pp.
This book was written for people who find the scientific process a compelling path to discovery but remain unconvinced by the Christian message. Polkinghorne explains that his method of inquiry entails making observations about reality, then seeking the best expla-

nations for these observations. This is the essence of scientific methodology, he asserts, but it is also the proper method of seeking truth in theology. He applies his method to the problems of creation, God's interaction with the world, the resurrection of Jesus, and the future of humankind. In this short collection of lectures Polkinghorne introduces ideas developed more fully in his earlier works.

_____. *Quarks, Chaos and Christianity: Questions to Science and Religion.* New York: Crossroad, 1996; 102 pp.
In this book, Polkinghorne takes issue with the popular view that science and religion are at odds. He believes, instead, that they function as friends: both involve the search for truth. He examines the notions of experiment and theory, and fact and interpretation in relation to both approaches to truth. He considers questions about the existence of God, how God creates, reductionism, miracles, and the resurrection, and examines how scientists can develop a reasonable faith. This volume provides a brief summary of six of his previous books.

Poole, Michael W. *Guide to Science and Belief.* Oxford, England: Lion, 1990; 128 pp.
Poole is a lecturer in science education at King's College, London. This little book is an attractively illustrated, well-written introduction to issues of science and faith for the general public. Poole considers a wide range of topics: how to understand science and scripture, the nature of explanation in both science and theology, the relationship between faith and scientific evidence, the nature of miracles, Galileo's troubles with the church over his Copernican views, biological evolution and the biblical concept of creation, and design and chance. As a Christian, the author takes a high view of both science and faith and seeks to integrate these two aspects of experience in an intelligent, informed, and carefully reasoned fashion.

_____. *Beliefs and Values in Science Education.* Buckingham, England: Open University Press, 1995; 130 pp.
Poole's purpose for this book is to promote the view "that science should be neither deified, denigrated, nor forced into demise," and to help science educators see "how spiritual, moral, social and cultural factors affect science." Poole first addresses the question of values in science education. He notes that some important aspects of reality are not open to scientific investigation. He is critical of logical positiv-

ism, which he says leads to "*scientism* not to science." He favors, instead, the perspective of "critical realism" which views scientific theories as interpretations of reality, not reality itself. He then discusses the role of belief in the development of scientific models, and the importance of both scientific models and models of the scientific enterprise itself to the teaching process. The final chapters of the book deal with environmentalism, cosmology, the Galileo affair, and Darwinism. Poole is critical of the view that mechanistic explanations for natural processes necessarily exclude the notion of ultimate plan and purpose, and he calls for a distinction between "evolution" as a biological theory and "evolutionism" as a philosophical worldview.

_____, and G. J. Wenham. *Creation or Evolution: A False Antithesis?* Oxford: Latimer House, 1987; 84 pp.
Poole and Wenham contend that both creation and evolution are consistent with scripture. They provide a description of evolutionary theory, an evolutionary interpretation of Genesis 1-3, and a critique of common creationist positions. They suggest that the notion of an immediate creation is based on an interpretation of Scripture not necessarily warranted by the text, and that Genesis tells us why creation occurred but not how. The authors clarify terms that are frequently misunderstood and misused; for example, evolution as process versus evolution as mechanism, and the processes of creation and evolution versus the philosophies of creationism and evolutionism. This book was written and published for members of the Church of England.

Postman, Neil. *Technopoly: The Surrender of Culture to Technology.* New York: Knopf, 1992; 222 pp.
Postman argues that the United States has become a technopoly, a society ruled by technology. This, he claims, has resulted in a loss of purpose, meaning, or hope. He posits that religion has served as the greatest source of hope to people through time, and that Genesis was a particularly important religious narrative in this respect. The loss of faith in Genesis and the rest of the Bible means that more and more people "make no moral decisions, only practical ones." By contrast, people who keep the biblical commandments "love God and express their love for Him through love, mercy, and justice to our fellow humans." Postman also writes "that the Biblical version of creation,

to the astonishment of everyone except possibly a fundamentalist, has turned out to be a near-perfect blend of artistic imagination and scientific intuition." He believes that America's public educational system is, in part, responsible for technopoly. Postman is not optimistic that Americans will regain their sense of meaning.

Price, Barry. *The Creation Science Controversy.* Sidney, Australia: Millennium Books, 1990; 244 pp.
The author is a Roman Catholic science teacher from Australia. His purpose is to debunk scientific creationism for anyone tempted to take its seriously. He does this by refuting creationist claims regarding thermodynamics, flood geology, fossils, dinosaurs, and many other topics. He reviews the controversies over the inclusion of creationist perspectives in science textbooks, as well as the court cases over the teaching of creationism in public schools. He examines the personalites of the scientific creationist movement, particularly Henry Morris and Duane Gish, and questions their integrity. He also details aspects of the creation science movement in Australisa. Price is critical of fundamentalist interpretations of scripture, believing that the Genesis creation accounts were derived from Babylonian myths.

Price, Zane H. *Atom to Adam: A Model of Theistic Evolution.* Los Angeles, CA: Probe Books, 1996; 115 pp.
Price is a retired research scientist from the University of California, Los Angeles, School of Medicine. He introduces his book by surveying the history of attempts to understand the relationship between the Bible and the "Great Book of Nature." He notes that early Christians saw no conflict between the two sources, but when nature became better understood starting in the seventeenth century, conflicts arose and continue to arise today. The writers of Genesis could not possibly have known what we know today, Price notes, so we cannot possibly think of Genesis as communicating science. Instead, he suggests that a "revelation of moral and religious truth" was its ultimate purpose. In Part One, "Atom to Earth," Price surveys the development of the physical universe, whereas in Part Two, "Earth to Adam," he addresses the history of life. He describes and accepts standard evolutionary models for the history of the universe and life, although at the same time he believes that a divine creator was behind it all. A fairly extensive glossary of scientific terminology is provided.

Pun, Pattle P. T. *Evolution: Nature and Scripture in Conflict?* Grand Rapids, MI: Zondervan, 1982; 336 pp.
Pun believes that Christians must evaluate information from both nature and the Bible as they ponder the issue of evolution. The first section of his book examines the scientific evidence favoring evolutionary theory. Here Pun summarizes evidences from both the earth sciences and biology, concluding that the theory of evolution has both strengths and weaknesses. Its primary strength involves its ability to describe microevolutionary processes at the population level. Its primary weakness is its capacity to explain macroevolutionary processes. The second section of the book deals with information from the Bible. Pun compares three Christian approaches to the history of life: fiat creationism, progressive creationism, and theistic evolutionism. On weighing all the evidence, Pun favors progressive creationism, although he agrees that this approach is not without its difficulties. He also favors a local flood view, which to him is a more credible position than held by universal-flood advocates. Pun concludes with a discussion of the dangers of philosophical evolutionism, a view that seeks to explain all reality in evolutionary terms.

Rae, Murray, Hilary Regan, and John Stenhouse (eds.). *Science and Theology.* Grand Rapids, MI: Eerdmans, 1994; 251 pp.
In this book, six scientists and theologians provide chapters on natural theology, arguments for the existence of God, what theology can learn from science, God and natural order, relativity and theology, and creation and causality. Each chapter is followed by comments from two respondents, a format that enhances the value of the essays. The authors believe that theology and science can work together, and they write from a generally Christian perspective.

Ratzsch, Del. *Philosophy of Science: The Natural Sciences in Christian Perspective.* Downers Grove, IL: InterVarsity Press, 1986; 165 pp.
Ratzsch, a philosopher of science at Calvin College, provides a carefully written work that introduces readers to the nature of science from a well-informed philosophical and Christian perspective. He explores the shift from empiricism, objectivism, and rationalism that dominated scientific thought from the mid-seventeenth century to the mid-twentieth century, to the paradigm-governed science described by Thomas Kuhn during the late twentieth century. Ratzsch criticizes the

extremes of both positions. He believes that science has a limited sphere of usefulness—that it has nothing to say about the ultimate origin and meaning of existence. Ratzsch also provides carefully reasoned apologies in response to so-called "scientific" challenges to religious belief. In the last two chapters he defends the legitimacy of science, examines the unique contributions that Christians can make to the scientific endeavor, and discusses the relevance of the Bible to one's understanding of science. Ultimately, Ratzsch argues, "the Christian has a broader context for thinking about science" than do naturalistic thinkers.

_____. *The Battle of Beginnings: Why Neither Side Is Winning the Creation-Evolution Debate.* Downers Grove, IL: InterVarsity Press, 1996; 272 pp.

Ratzsch provides an insightful analysis of arguments used by creationists and evolutionists in their debate. This is not a book about interpretation of data but about sound logic. Ratzsch briefly reviews the history of geology and biology prior to publication of Darwin's *Origin of Species.* He notes that Darwin convinced most people that some type of descent with change had occurred in the history of life. Ratzsch shows that creationists frequently misunderstand Darwin's theory, thus reducing the effectiveness of their arguments against evolutionism. He suggests that when creationists use the term *evolution,* they often fail to distinguish between the biological theory of evolution and evolutionary philosophy. This failure to communicate clearly further erodes the effectiveness of the creationist position. Ratzsch then turns his attention to evolutionist misunderstandings about creationism and shows how these misunderstandings lead to irrelevant arguments. He also examines arguments favoring and disfavoring theistic evolution and explains how some of these arguments are flawed. His survey of the nature and philosophy of science suggests that science is subject to serious limitations. These limitations, according to Ratzsch, should lead to humility on the part of both evolutionists and creationists as they collect and interpret data and shape their theories. This outstanding book probes the very heart of the creation/evolution controversy.

Ross, Hugh. *Creation and Time: A Biblical and Scientific Perspective on the Creation-Date Controversy.* Colorado Springs, CO: NavPress, 1994; 187 pp.

Ross, a Christian astronomer and former post-doctoral fellow at the California Institute of Technology, examines the contentious age-of-the-universe debate in this book. He first reviews the past two thousand years of this debate, then, using internal biblical evidence, argues that one cannot use the Bible to defend a young universe position. Revelation, he says, comes from both the Bible and nature, and the revelation from nature is clear—the universe is very old. Ross arrives at this conclusion from a variety of evidences, including rates of stellar decay, abundances of radioactive elements, and the expansion of the universe. He provides an extensive discussion of the problem of the occurrence of death before the entrance of sin, a major concern of many conservative Christians when faced with evidence from the fossil record. Ross suggests that it is absurd for Christians to argue over the age of the universe. The real issue, he says, is the infinitesimally small probability that life arose spontaneously, regardless of the length of time available.

Russell, Colin A. *Cross-Currents: Interactions Between Science and Faith.* Grand Rapids, MI: Eerdmans, 1985; 272 pp.
A distinguished science historian at England's Open University, Russell takes a holistic view of science and faith, believing that these two aspects of human experience are not necessarily in conflict. Russell discusses the history of science and faith from the sixteenth to the twentieth century, including discourses on the Copernican revolution, Puritans and science, natural theology, the rise of flood geology, Darwinian evolution, Romanticism and nineteenth century science, modern physics and theology, and contemporary Christian attitudes toward the environment. Ultimately, Russell attempts to demonstrate to students that Christianity and science interact with one another in important and positive ways.

Russell, Robert John, William R. Stoeger, S. J., and George V. Coyne, S. J (eds.). *Physics, Philosophy, and Theology: A Common Quest for Understanding.* Notre Dame, IN: Notre Dame Press, 1989; 408 pp.
This book is the outgrowth of sessions held at the Papal Summer Residence at Castel Gandolfo and at the Center for Theology and the Natural Sciences in Berkeley. It includes a message from Pope John Paul II who suggests that "Science can purify religion from error and superstition; religion can purify science from idolatry and false abso-

lutes." Other authors deal with historical and contemporary relations in science and religion, epistemology and methodology, and contemporary physics and cosmology in relation to philosophy and theology. Several authors warn that religion should never be tied to a particular scientific paradigm. One essay posits that "Fundamentalism is not as is sometimes supposed, an anachronistically surviving precursor of modern rationalism, but a byproduct of it." The authors present a wide range of viewpoints, all seeking to foster a mature, informed approach to issues of science and faith.

Sáenz, Braulio A. *To God Through Science.* New York: Vantage, 1984; 53 pp.
In this small, philosophical treatise, Sáenz argues for the existence of God on the basis of of intelligence and reasoning. Sáenz distinguishes between concrete and abstract reasoning—the former involves responding to current reality, while the latter involves the ability to foresee and plan for future events. Sáenz posits that, while all animals show at least minimal levels of concrete intelligence, only humans exhibit abstract intelligence. He believes that even individual cells exhibit amazing levels of intelligence in their ability to adjust to changing circumstances. This universality of intelligence, from cells to humans, Sáenz writes, provides evidence for the existence of a supreme intelligence, God.

Sailhamer, John H. *Genesis Unbound: A Provocative New Look at the Creation Account.* Sisters, OR: Multnomah Books, 1996; 257 pp.
Sailhamer, who teaches Old Testament at Western Seminary and at Northwestern College, takes issue with three common assumptions regarding the Genesis creation account: (1) that the purpose of this account is to describe *how* God created, (2) that Genesis 1:2 refers to a formless void from which God shaped the earth, and (3) that this account addresses the formation of the world as a whole. Instead, Sailhamer posits that the Genesis story seeks only to establish the fact of God's creatorship, a process that may have extended over billions of years, and to report the preparation of that part of the earth used by Abraham and his descendants. He believes that the Genesis creation text must be considered on its own merits and not through the eyes of science. Moreover, he suggests that this text must be studied within the context of the entire Pentateuch if it is to be properly understood. This approach, which Sailhamer calls historical creationism, provides

a biblical view of origins which, he believes, harmonizes with contemporary scientific perspectives on the history of the universe.

Samuelson, Norbert M. *The First Seven Days: A Philosophical Commentary on the Creation of Genesis.* Atlanta, GA: Scholars, 1992; 186 pp.
Samuelson writes from a rabbinic tradition with a personal commitment to liberal Judaism. In this book he provides "an in-depth, logical analysis of the concept of creation contained in the original Hebrew text of the first thirty-four verses of the Book of Genesis." This consists of a "word by word, and sentence by sentence" evaluation of the masoretic Hebrew text. He tries to let the "text speak for itself" in an effort to determine its original meaning. Samuelson suggests that the text allows for a number of possible interpretations, including those of modern cosmologists based in contemporary physics. By contrast, he says the text precludes the interpretations of Christian fundamentalists who read the text literally. He concludes with a "description of the concept of the origin and general nature of the universe" assumed by pre-rabbinic Judaism and "critical judgment about what is true about the conception of the origin and nature of the universe in the Hebrew Scriptures."

_____. *Judaism and the Doctrine of Creation.* New York: Cambridge University Press, 1994; 362 pp.
Samuelson considers the relationship between modern Jewish philosophy and contemporary science, particularly in relation to the origin and nature of the universe. His analysis centers on a book by Franz Rosenzweig, who, during the nineteenth century, provided one of the best modern works on creation in Jewish philosophy. Samuelson first analyzes Rosenzweig's book, then demonstrates that it is representative of a standard Jewish approach to Genesis. This approach sees God, through an act of will, creating out of nothing an eternal universe with intrinsic meaning and moral value. He then asks if this Jewish perspective is compatible with contemporary philosophy and physics, and answers that it does. In fact, Samuelson believes that much of traditional Jewish thought harmonizes with contemporary philosophy and science.

Schroeder, Gerald L. *Genesis and the Big Bang: The Discovery of Harmony Between Modern Science and the Bible.* New York: Bantam,

1992; 212 pp.
Believing that conflicts between science and religion result from misinterpretations of scripture, Schroeder, a physicist, goes far beyond a literal reading of the biblical text. He does this by an unusual blend of conservative creationism and Jewish philosophy, melded together with modern cosmology. For example, he believes that the six day creation and the Big Bang, which occurred fifteen billion years ago, were both literal events. This is possible because the creation occurred in God's time, which is relativistic. He also believes that Adam, who lived around 3800 B.C., was the first human to take on God's image, even though humans had been around for 300,000 years. He holds that life could not have arisen by random events and that the fossil record does not agree with the theory of gradualistic evolution. This is an idiosyncratic but thought-provoking book.

Seely, Paul H. *Inerrant Wisdom: Science and Inerrancy in Biblical Perspective.* Portland, OR: Evangelical Reform, 1989; 216 pp.
Seely, a biblical scholar, takes the interesting religious viewpoint that the Bible is "inerrantly wise." He completely rejects the fundamentalist notion of inerrancy which holds that everything in the Bible is factually true, not only in matters of faith, but also in matters of biology, geology, astronomy, history, and so on. Seely believes that his inerrant wisdom view allows Christians to accept the authority of scripture without bending reality into a literal reading. He argues that inerrancy is not taught by the Bible itself, then shows that many historical and scientific details of the biblical text are inaccurate. Seely believes that one can accept the view that the Bible is inspired without trying to explain all of these errors away.

Sharpe, Kevin J. *From Science to an Adequate Mythology.* Auckland, New Zealand: Interface, 1984; 105 pp.
Sharpe, who has doctorates in both religious studies and mathematics, is a chaplain at the University of Auckland. He believes that, by themselves, science and both liberal Christianity and conservative Christianity are inadequate sources of values in today's society. He argues that a new "mythology" is needed, one "founded on the integration of secular-scientific knowledge and that of the Christian religion in which neither is subordinated to the other." While he believes that the Genesis creation "stories are meant to be taken literally," conservative Christianity "is totally separated from our secular life,

and therefore cannot influence it." But Sharpe disavows Christian liberalism as well. He calls, instead, for a new form of liberal Christianity, one that is "firm and strong in its theology without the gaps and confusion of its predecessor." He envisions science and theology as forming the two rails of a ladder, with the connecting rungs representing shared viewpoints. He would like to see theology derive insights from science and science derive insights from theology.

———. *David Bohm's World: New Physics and New Religion.* Lewisburg, PA: Bucknell University Press, 1993; 168 pp.
In this book Sharpe discusses the mathematics, physics, and metaphysics of the Jewish phyicist David Bohm in relation to Christianity. As Sharpe points out, Bohm emphasizes the unity of the cosmos, that what happens in one part of the creation may affect the entire creation. Recognition of this fact is the only way we can understand reality. Sharpe believes that there are various levels of reality at which religion and science interact. He uses the metaphor of a ladder. The ladder stands on the ground of the real world. One rail of the ladder represents science, while the other represents theology. The rungs represent points of intersection between the two disciplines. This book provides an interesting study of the relationship between science and religion in the context of the physical sciences.

Sitchin, Zecharia. *Genesis Revisited: Is Modern Science Catching Up with Ancient Knowledge?* Santa Fe, NM: Bear, 1991; 343 pp.
Sitchin, a specialist in economic history and reader of the ancient Sumerian language, believes that Genesis provides the highest possible level of scientific knowledge. He supports his thesis by appealing to unique interpretations of Near Eastern creation myths which he uses to validate the biblical text. He provides an entertaining, although fantastic, set of assertions. Some examples: advanced beings visited earth to collect gold, found that it was difficult to mine, and decided instead to produce the first humans by genetic engineering; the Tower of Babel was a launch pad for a space rocket; and the Sumerian view that the planets are alive is supported today by the Gaia hypothesis. This bizarre book is poorly referenced but skillfully written.

Smith, Wolfgang. *Cosmos and Transcendence: Breaking Through the Barrier of Scientistic Belief.* LaSalle, IL: Sherwood Sugden, 1985; 168 pp.

Smith is a Christian mathematician. In this book he differentiates between scientific knowledge and scientistic belief. He suggests that many scientists have an inadequate understanding of the philosophy of science. He notes that scientific models only approximate reality. Smith believes that God is the ultimate cause of all things and that nature is a reflection of God's face; moreover, eternity is "in the ever-present 'now': like the Kingdom of God, it lies 'within.'" He remains open to the possibility of theistic evolution but points out the seeming explosion of life implied by the Cambrian fauna. Smith is critical of psychoanalysis and writes that Freudian theory is an inversion of the Christian truth and that Jungian views are rooted in Gnostic error. He opposes the notion of progress and posits that macroevolutionary theory has been spiritually damaging. According to Smith, humans today have "become sophisticated, skeptical and profane."

Spanner, Douglas. *Biblical Creation and the Theory of Evolution.* Exeter, England: Paternoster Press, 1987; 191 pp.
Spanner, formerly chair of plant biophysics at the University of London and now a minister in the Church of England, believes that scientists commonly overlook meaning in their search for mechanism, while clergy, on the other hand, often lose their confidence in the Bible due to a misunderstanding of science. Spanner has a high regard for scripture and affirms his belief that God is continually involved in sustaining his creation. He also respects the scientific enterprise and believes that God works through the normal processes of nature which are open to scientific exploration. He believes that the "'chance' element in evolutionary theory presents no insuperable obstacle to biblical doctrine." He deals with a variety of specific issues in the main text and appendices, including the notion of contingency, the anthropic principle, the origin of life, Darwinism, the days of creation, providence, the flood, and the problem of evil.

Sproul, R. C. *Not a Chance: The Myth of Chance in Modern Science and Cosmology.* Grand Rapids, MI: Baker, 1994; 234 pp.
Sproul is a Christian apologist and professor at Reformed Theological Seminary, Orlando. In this book he attempts to show that belief in God can be a rational response to reality. He criticizes scientists who posit that the universe came about as a result of chance. Sproul notes that chance refers to mathematical probabilities which can do nothing. Chance is merely a concept with no ability to act. He be-

lieves there are only three possible ways to view reality: matter and energy arose from nothing, matter and energy are eternal, or God is the source of all that is. To Sproul, the last of these three options makes the most rational sense.

Steward, Don. *The Creation.* San Bernardino, CA: Here's Life, 1984; 178 pp.

This book, adapted from a 1980 Dutch publication, is published as part of the "Family Handbooks of Christian Knowledge" series from Campus Crusade for Christ. Steward argues from the viewpoint of scientific creationism and favors a young earth model for earth history. The book is attractively illustrated and contains a good discussion of the meanings of terms used by scientists such as microevolution, macroevolution, and species. Treatment of topics such a mutation and population growth, however, belies serious deficiencies in the author's understanding of biological science. He relies heavily on the arguments and authority of other scientific creationists such as Henry Morris and John Whitcomb.

Stoner, Don. *A New Look at an Old Earth.* Paramount, CA: Schroeder, 1992; 192 pp.

Stoner believes that young earth creationism serves as a stumbling block to Christian evangelism and prefers to read Genesis from a day-age point of view. He rejects the notion of a literal six-day creation on the basis of several arguments and interpretations: the "evening and morning" of Genesis probably means "from beginning to end"; the "numerical adjectives" in Genesis 1 should be interpreted in the same non-literal way as "last" in the term "last day" used in reference to the end times; biblical texts refer to God as eternal, implying that his earth must also be very old; the fact that we can see distant stars proves that they were present long ago; and the universe looks very old, therefore it must very old, for God does not lie. In line with this last assertion Stoner suggests that Adam "might have been created as a baby or even as an embryo," rather than with the deceptive appearance of age. He suspects that God wanted people to discover the Big Bang, progressive creation, and long ages for themselves.

Taylor, Ian T. *In the Minds of Men.* Toronto, Canada: TFE, 1992; 498 pp.

Taylor attempts in this book to provide a comprehensive survey of creationist views. Following an examination of the views of Socrates, Plato, and Aristotle, Taylor describes the events leading to the publication of Charles Darwin's *Origin of Species* in 1859. He discusses the influence of Malthus, Lyell, and Wallace on Darwin's views. Taylor points out that Darwin and his supporters believed many transitional forms of organisms eventually would be found as fossils. This never happened, says Taylor, providing one of the most vexing problems for evolutionists. Taylor recounts creationist arguments for a young earth and solar system, including the small amounts of meteoritic dust on the earth and moon, the low concentration of helium in the earth's atmosphere, rapidly forming cave deposits, and polonium halos. He debunks evolutionary theory in view of Ernst Haeckel's drawings of embryos distorted to favor an evolutionary interpretation of development, the Piltdown hoax, and the Java man deception. Today, evolutionism influences all areas of study, he observes, and it is responsible for the evils of Marxism, Hitler's eugenic views, and secular humanism.

Templeton, John Marks. *The God Who Would Be Known: Revelations of the Divine in Contemporary Science.* San Francisco: Harper & Row, 1989; 412 pp.
In this book Templeton, the Christian investment mogul, notes that the uncertainty principle and the anthropic principle allow for more open, often more theistic, views of nature than were common when Newtonian physics dominated the scientific scene. He believes that these newer paradigms are leading to less hostile interactions between people of science and people of faith. He finds evidence for faith in, among other topics, astronomy, the infinitesimal, genetics, the organizing forces of the universe, and the history of humanity. The author views God as one who upholds the universe with continual activity, without which there would be nothing.

_____ (ed.). *Evidence of Purpose: Scientists Discover the Creator.* New York: Continuum, 1994; 212 pp.
In this anthology volume, Templeton brings together essays by ten distinguished scientists: Paul Davies, John Eccles, Owen Gingerich, Walter Hearn, Daniel Osmond, Arthur Peacocke, John Polkinghorne, Robert John Russell, Russell Stannard, and David Wilcox. Although the contributors express divergent philosophical perspectives, each of the writers is struck by evidence from the natural world that seems to

point in the direction of design and purpose. The authors reject the notion that the practice of science forces one to view the universe from a naturalistic point of view. Indeed, it is suggested that the very fact that science and mathematics work gives evidence of design. This book represents part of Templeton's continuing broad-scale effort to bring the concepts of design and purpose back into the intellectual marketplace.

_____, and Robert Herrmann. *Is God the Only Reality?* New York: Continuum, 1994; 190 pp.
Templeton, a global investor with a keen interest in science and the Christian faith, and Herrmann, a professor of chemistry at Gordon College, argue that recent findings in science point toward "a deeper meaning of the universe." The authors point out that recent discoveries in anthropology, biology, cosmology, and physics show that natural systems exhibit remarkably greater levels of complexity than expected. This complexity, they believe, points to the existence of God and purpose in nature. While the tone of this book is not "preachy," it does call for a reconsideration of the concept of God on the part of scientists. It also hopes that more effective communication between theologians and scientists can be developed.

Thomas, J. D. (ed.). *Evolution and Faith.* Abilene, TX: Abilene Christian University Press, 1988; 232 pp.
This collection of essays, primarily by faculty members at Abilene Christian University, was written to "help lift the veil of ignorance about evolution and Biblical creation." The authors believe that the Genesis account of creation should be read as "a straightforward, sober statement of what actually happened." The authors do not find the notion of a several-billion-year-old-earth objectionable, accept the theory of an expanding universe, and do not insist that the Genesis flood was universal. They do reject, however, organic evolution, theistic evolution, progressive creation, concordism, anti-scientism, fideism, the gap theory, the day-age theory, and creation science. The authors thus present an idiosyncratic position not easily characterized, but one that underscores the incredible diversity of opinion on the topic of earth history in relation to religious faith.

Thomson, Alexander. *Tradition and Authority in Science and Theology.* Edinburgh: Scottish Academic Press, 1987; 108 pp.

Thomson, who is a biochemist and a theologian, first demonstrates that scientists, like religious people, exhibit faith—faith that the world makes sense and that the scientific community provides the best understanding of reality. He also shows that authority and tradition play important roles in both science and the Reformed Church, for which he writes. Thomson's views are based in part on Michael Polanyi's philosophy of science. This philosophy, Thomson believes, guards against two common extremes among Christians who attempt to interpret reality: the Protestant "Scripture only" position, and the Catholic view that Scripture and tradition should be taken as equal authorities.

Tiffin, Lee. *Creationism's Upside-Down Pyramid: How Science Refutes Fundamentalism.* Amherst, NY: Prometheus Books, 1994; 229 pp.
In this book, Tiffin, a pastor, takes aim at scientific creationism primarily from scientific, not biblical, grounds. In his first section Tiffin discusses the assumptions and methodologies of creationism and finds them wanting. In the second section he examines the scientific arguments of creationism, concluding they lack supporting evidence. Finally, he deals with public concerns over creationism. Tiffin is particularly concerned with the issue of flood geology and endeavors to show the implausibility of a literal worldwide flood. He agrees with his opponents that such an event could only be the result of miraculous intervention but disagrees that the geological evidence points in that direction.

Torrance, T. F. *Reality and Scientific Theology.* Dover, NH: Longwood, 1985; 212 pp.
Torrence is a theologian who views both science and theology as legitimate and important sources of truth in modern culture. He believes that both disciplines search for the same objective, God-based reality—theology seeks to know God, whereas science seeks to know God's creation. He is impressed by the success of modern physical science, which, he says, arose from biblical roots toward the end of the medieval period. He is very critical of existential and operational influences on all aspects of contemporary culture, including science and theology. It is the role of Christian theology, he contends, to develop a "scientific theology" that avoids these and other destructive, cultural influences. He believes that scientific theology should be

marked by the same level of intellectual rigor as characterizes modern physical science. The six chapters of this book make for difficult but thought-provoking reading.

Van Till, Howard J. *The Fourth Day: What the Bible and the Heavens Are Telling Us About the Creation.* Grand Rapids, MI: Eerdmans, 1986; 280 pp.
Van Till is a professor of physics and astronomy at Calvin College and a frequent writer on issues of science and faith. In this book he first examines biblical views of reality. If we want to take the Bible seriously, he writes, we must, among other things, affirm its true status, character, and function. This includes understanding its covenantal structure; seeking the original meaning for each of its messages; recognizing that it reveals God as the originator, sustainer, and governor of nature; and assuming its nonscientific genre. Secondly, Van Till examines the nature of science and its proper domain. This domain, he believes, is restricted to explanations of correlations between properties and behavior of the physical universe. He discusses in some detail various scientific discoveries and theories, and he concludes that "cosmic history is evolutionary in character." In his third and final section, Van Till integrates the biblical and scientific views by suggesting that questions about the "physical properties, material behavior, and temporal development of the universe" are properly addressed by science, whereas queries "about the status, origin, governance, value, and purpose of the universe" are properly addressed by scripture. He believes that "creation" and "evolution" refer to entirely different classes of events and asserts that the real debate is "between atheistic naturalism and biblical theism."

_____, Robert E. Snow, John H. Stek, and Davis A. Young. *Portraits of Creation: Biblical and Scientific Perspectives on the World's Formation.* Grand Rapids, MI: Eerdmans, 1990; 189 pp.
The authors of this book have all been associated with the Calvin Center for Christian Scholarship. They depict young earth creationists as zealous and somewhat careless interpreters of scripture, and scientific creationism as a distortion of the scientific process. Several claims of scientific creationists, such as the so-called "missing layers" of rock in the Grand Canyon, are carefully and systematically dismantled. The authors discuss what they believe to be the necessary requirements for good science—competence, integrity, and sound

judgment. They also examine what they believe the Bible says about God as creator and sustainer. The authors support standard scientific models regarding the formation of the universe and the earth, as well as an evolutionary history of life, albeit from a strongly theistic perspective.

_____, Davis A. Young, and Clarence Menninga. *Science Held Hostage: What's Wrong with Creation Science AND Evolutionism?* Downers Grove, IL: InterVarsity Press, 1988; 189 pp.
All three authors are at Calvin College, where Van Till teaches astronomy and Young and Menninga teach geology. These writers have become well known for their attempt to delineate appropriate provinces for science and theology. This book continues that effort. Part 1 discusses how authentic science operates. Here the stage is set for the rest of the book with the thesis statement: "Science held hostage by any ideology or belief system, whether naturalistic or theistic, can no longer function effectively to gain knowledge of the physical universe." Parts 2 and 3, respectively, examine case studies revealing how science is held hostage by some of the claims of scientific creationists and atheistic naturalists. Whatever one's philosophical perspective, this book makes for profitable reading.

von Fange, Erich A. *Noah to Abram: The Turbulent Years.* Syracuse, IN: Living Word Services, 1994; 372 pp.
This book examines biblical history between the time of Noah and Abraham. The author provides an unusual creationist reconstruction of the world before the time of the flood. During these antediluvian years, says von Fange, the world "was full of deadly peril," including "great catastrophic events like "meteoric showers, huge strikes from outer space, . . . plagues, . . . massive floods, earthquakes, and volcanic eruptions." Moreover, after the flood "Nature has gone on many a rampage," he opines. "We can see the evidence in the scarred earth, and every year there are sobering tales of floods, hurricanes, earthquakes, volcanic eruptions, [and] giant mudslides." He describes human history from the departure of Noah from the ark, through to the time of Abraham and the destruction of Sodom and Gomorrah. Von Fange provides young earth, creationist interpretations of cave men, the Ice Age, the Stone Age, technology and invention, language, and climatology. A chart containing a time line of biblical history is provided.

Weaver, John David. *In the Beginning God: Modern Science and the Christian Doctrine of Creation.* Macon, GA: Smyth and Helwys, 1994; 209 pp.

Weaver was a senior lecturer in geology at Derby College and a Baptist pastor; he became a fellow and tutor in pastoral theology at Regent's Park College, Oxford. He wishes to remove science as a hindrance to people's acceptance of Christ, to help Christians who are fearful of science for what it might do to their faith, and to provide an explanation for suffering in the world. He discusses evidence for purpose and design in physics, biology, and geology. He believes that we are limited in our ability to discover God through nature. He is critical of the fundamentalist view that humans were created separately and sees them instead as products of evolution. Suffering, he writes, is caused by evolutionary processes, and God suffers when we suffer because he journeys with us as we evolve. Weaver also discusses the interpretation of Genesis, the interaction between science and faith, and the role of apologetics in preaching.

Whitcomb, John C. *The Early Earth: An Introduction to Biblical Creationism.* Grand Rapids, MI: Baker, 1986; 174 pp.

Whitcomb, coauthor with Henry Morris of *The Genesis Flood,* is a conservative professor of theology and Old Testament at Grace Theologial Seminary. He believes that everyone has a simple choice when confronted with the issue of origins: one can "put his trust in the written Word of the personal and living God who was there when it all happened, or else [put] his trust in the ability of the human intellect." Whitcomb believes the first option requires one to accept the Genesis creation story as historically and scientifically true in every detail. He believes that other approaches, such as the day-age theory, the ruin-reconstruction theory, theistic evolution, and all other attempts to better accommodate Genesis to contemporary scientific models can lead to a loss of confidence in the entire Old Testament. This book provides a good introduction to the fundamentalist background of scientific creationism.

_____. *The World That Perished.* Grand Rapids, MI: Baker, 1988; 178 pp.

Whitcomb provides explanations for various aspects of flood geology from his literalist interpretive framework for the Bible. Whitcomb does not view his position, however, as interpretive. He believes that

Christians without literalist views have simply succumbed to pressure from evolutionary geologists. Whitcomb discusses issues such as the redistribution of organisms after the flood, the presumed extinction of dinosaurs, and the universality of the Genesis flood. While his views range far outside the perimeter of normative science, Whitcomb believes that the Bible's authority must be subject to verification by the results of scientific research. Without this "scientific control," Whitcomb suggests, the Bible's "concepts become as puerile and insipid as the adventures of ancient Babylonian deities." Presumably, the science that Whitcomb has in mind is creation science.

White, A. J. Monty. *Why I Believe in Creation.* Durham, England: Evangelical Press, 1994; 24 pp.
White is a chemist and a popular writer and speaker on the topic of creation. This booklet provides a brief synopsis of conservative creationism. White notes that the Bible contains not even the slightest suggestion that evolution occurred. He discusses the laws of thermodynamics and posits that these laws are assumed by Genesis 1:1. He is critical of the standard evolutionary interpretations of Stanley Miller's origin of life experiment and says that his conclusions are invalidated, regardless of whether one assumes an oxydizing or a reducing atmosphere for the primitive earth. White also briefly discusses the origin of the universe, the appearance of life and humankind, and fossils. He includes a list of creationist organizations at the end.

Whitehouse, W. A. *Creation, Science, and Theology: Essays in Response to Karl Barth,* Ann Loades, editor. Grand Rapids, MI: Eerdmans, 1981; 247 pp.
In this book Whitehouse, a British theologian, examines the perspectives of Karl Barth, the prominent Swiss theologian. He takes up Barth's views on creation, the nature of man, divine providence, and the relationship between theology and natural science. Whitehouse contends that one must look for a correspondence between Christian and non-Christian worldviews in the search for truth. In accord with Barth, Whitehouse believes that meaningful revelation occurs through an encounter with Jesus and a reflection on the created world. Moreover, because all humans are God's creation, they are automatically affirmed, chosen, and accepted by God. Whitehouse notes that while today's theologians and scientists live in somewhat different concep-

tual worlds, science grew out of a theological worldview. Despite this historical connection, he writes, science obscures this indebtedness by explaining natural processes purely in terms of autonomous entities. Whitehouse suggests that by acknowledging the metaphysical assumptions of modern science, scientists and theologians could complement rather than compete with one another.

Wiester, John. *The Genesis Connection.* Nashville, TN: Thomas Nelson, 1983; 254 pp.
Wiester, a cattle rancher and retired president of a scientific instrument company, has degrees in geology and business administration. He became interested in science and faith issues after his recent conversion to Christianity. Wiester believes that the issue is not one of creation versus evolution, but of creator versus no creator. He suggests that both "modern science and the Bible are presenting the same record of historical events," and he uses the days of creation as a framework for his topics. He believes that the six days of creation in Genesis 1 represent six ages in the history of the universe and otherwise accepts the standard scientific view of cosmic and geologic time. According to Wiester, God was involved in the unfolding of creation all along. He thinks it is likely that God may have used the evolutionary process to create diversity of life-forms but does not rule out the possibility of sudden and miraculous interventions from time to time. Humans, he believes, were directly created by God, although he does not ignore the fossil hominid record, which he considers to be equivocal. This nicely illustrated book closes with an appendix on scientific dating techniques and a glossary.

Wilkinson, David, and Rob Frost. *Thinking Clearly About God and Science.* East Sussex, England: Monarch, 1996; 206 pp.
Scientist Wilkinson and theologian Frost believe that all truth originates with God and that science and Christianity are concordant means through which meaning and truth are discovered. Frost discusses various arguments favoring the existence of God, concluding that neither science nor religion can prove that God exists, but that both point toward the existence of some type of creative intelligence. Wilkinson examines the limitations of both science and theology and notes that both use evidence to come to their respective conclusions. Consequently he believes they would do well to cooperate with one another in the search for truth. Frost considers the complexity of eth-

ical issues brought about by modern civilization, and posits that humans will benefit if science and religion cooperate in the construction of a good society. Wilkinson rejects the notion that scientific advancements in our understanding of the history of the universe preclude the need for a creator. While science may be able to describe what happened in the past, it cannot explain why these things happened, which is the prerogative of religion. He also rejects the view that scientific explanations rule out the possibility of miracles. Frost concludes the book by reaffirming the position that both scientists and Christians seek truth and that scientists can also be Christians.

Woodmorappe, John. *Noah's Ark: A Feasibility Study.* Santee, CA: Institute for Creation Research, 1996; 298 pp.
Woodmorappe analyzes the dimensions, capacities, occupants, and on-board management of Noah's ark, using numerical data from the Genesis text. He attempts to demonstrate mathematically and logically that the ark could, indeed, have carried out its assigned mission of saving representatives of all land-dwelling animals during a worldwide deluge. He assumes that the original Genesis "kinds" preserved on the ark represented modern-day genera which subsequently diversified. In view of this assumption Woodmorappe estimates that sixteen thousand animals were carried on the ark, with an average body mass of a small rat. At the larger end of the size range were the dinosaurs. He argues that eight intelligent people could have provided adequate food and fresh water for this menagerie and also kept their quarters clean. Woodmorappe also explains how Noah and his family could have provided adequate ventilation and lighting for their charges and posits how aquatic animals and land plants survived outside the ark. He believes that rapid speciation after the flood was responsible for the earth's post-flood repopulation and redistribution process.

Wright, John. *Designer Universe: Is Christianity Compatible With Modern Science?* Crowborough, England: Monarch, 1994; 158 pp.
Wright, a lay reader in the Anglican Church and retired Director of Health and Safety for Nuclear Electric in England, directs the "Science and Faith" program at Luton Industrial College. In this book he attemps to explain to the general reader how the findings of contemporary science fit with the teachings of Christianity. Wright examines the evidence for God as designer in three topic areas: cosmology, particle physics, and biology. He believes that the Big Bang was the

moment of creation and that the anthropic principle shows there is purpose in existence; that quantum behavior, relativity, and chaos interact to create a universe of subtlety and fruitfulness; and that God worked through Darwinian-type evolutionary processes to prepare the way for humankind, the highest creation. He rejects the notion that life itself evolved spontaneously. Wright also considers in this book the problems of design, pain and suffering, goodness, and miracles.

Youngblood, Ronald F. (ed.). *The Genesis Debate: Persistent Questions About Creation and the Flood.* Grand Rapids, MI: Baker, 1990; 250 pp.
Each of eleven questions involving creation and the flood are answered by essayists with opposing views. Some of the essayists are biblical scholars, while others are scientists. Many work at Christian institutions such as Liberty University, Luther Northwestern Seminary, American Baptist Seminary, the Institute for Creation Research, Dallas Theological Seminary, and Wheaton College. The questions include: Were the days of creation literal 24-hour days? What was the actual order of the events reported in Genesis 1? What is the age of the earth? Did God use evolution to create? Is the doctrine of the Trinity introduced by Genesis 1? Why was Cain's offering rejected by God? Were there pre-Adamic humans? Were the reported ages of the patriarchs before the flood real? Who were the sons of God mentioned in Genesis 6? Was Noah's flood a local event or worldwide in extent? Does Genesis 9 justify capital punishment? The opposing essays are printed in parallel for easy comparison.

NONTHEIST REFERENCES

Anderson, Walter Truett. *To Govern Evolution: Further Adventures of the Political Animal.* New York: Harcourt Brace Jovanovich, 1987; 376 pp.
Anderson, a political scientist, dismisses *both* Darwinism and theism in his first two chapters. He then sets about to warn readers that humans must themselves intervene in process of evolution, or else dire consequences will follow. In his final, and largest, section of the book Anderson details his agenda for developing the political muscle for governing evolution. He implores theists to abandon their ways and to join him in his new naturalistic religion. Anderson's icono-

clastic prescription for the future will strike many readers as Orwellian, but his book may engender some lively discussions.

Asimov, Isaac. *In the Beginning: Science Faces God in the Book of Genesis.* New York: Crown, 1981; 234 pp.
Asimov is a biochemist and a prolific popularizer of science. This book provides his verse-by-verse commentary on the book of Genesis. Asimov believes that Genesis was cobbled together from a variety of textual sources. He considers Genesis to be a wonderful success for the early Hebrew people, for it provides a "rational and . . . inspiring" story about the beginnings. He notes that, while both scientists and religionists argue over the interpretation of evidence in their respective areas of interest, only science settles on final truth through an allegiance to the weight of *compelling* evidence. He suggests that there is no scientific evidence for the existence of a divine being but neither is there is evidence that such a being does not exist. This book provides an interesting look at Genesis from the perspective of one of the twentieth century's bestknown scientific materialists.

_____. *Beginnings: The Story of Origins—Of Mankind, the Earth, the Universe.* New York: Walker, 1987; 271 pp.
Asimov presents a non-technical introduction to the evolution of humans, *backward* in time to the emergence of life, the formation of Planet Earth, and the development of the cosmos. He acknowledges that there is much we do not yet know about this history but confidently provides an outline of the process. He discusses the scientists who made the discoveries leading to our models of the past and describes their work. He discounts the notion that Ussher's biblical chronology is of any practical use today and argues that the Bible and evolution are incompatible. This is a well-written, popular exposition of the standard evolutionary approach to origins.

_____, George Zebrowski, and Martin Greenberg (eds.). *Creations: The Quest for Origins in Story and Science.* New York: Crown, 1983; 351 pp.
This volume is an entertaining collection of 27 short stories, articles, and book excerpts—some of it science fiction—by a variety of authors on the origin of the cosmos, the solar system, life, and humankind. Each of the four major sections begins with a quotation from Genesis. An essay and notes by Asimov express concern over the

threat of creationism, which he considers to be a religiously based superstition. He believes that creationist attacks against evolutionary theory are without content, and he expresses particular consternation over what he sees as the disastrous consequences of teaching creationism in the public schools.

Bailey, Lloyd R. *Genesis, Creation, and Creationism.* New York: Paulist Press, 1993; 259 pp.
Bailey is an associate professor of Hebrew Bible at Duke Divinity School. In this book he explores the meaning of the Genesis creation story in light of his understanding of the original text, and he compares this meaning with the interpretations of young earth creationists. Bailey believes that the Genesis story should be seen as a polemic against the polytheism common in Palestine when Genesis was written, and that to try to understand this text in scientific terms does an injustice to its intended purpose. Thus, he believes, scientific creationism is in conflict with not only mainstream science but also the biblical record it purports to defend. Bailey quotes approvingly from Carl Sagan and attempts to show compatibility between Sagan's naturalistic cosmology and the message of the biblical Genesis. Sixteen appendices deal with a variety of topics ranging from the cosmology of the ancient Semites to anti-evolutionism in today's court system.

Duschl, Richard A. *Restructuring Science Education: The Importance of Theories and Their Development.* New York: Teachers College Press, Columbia University, 1990; 155 pp.
Duschl is concerned about the state of science education today. He objects to several methods of science teaching. For example, he is opposed to the old method of giving students lists of scientific terms to memorize. He also objects to the teaching of science as if it was a completed body of knowledge. Moreover, he believes that in view of Thomas Kuhn's work on scientific revolutions, students should not be given the impression that science simply grows as new knowledge is added. He suggests that scientific theories can exist at different levels. For example, some theories, like the theory of evolution, are well established, even though they may contain some internal inconsistencies. Others, such as scientific creationism, are highly speculative but may someday be shown correct. According to Duschl, students should learn science in the same way that science is

done—through investigation and the replacement of old ideas with new ones. Duschl is clearly an educator, not a scientist, as a number of textual errors indicate. Nonetheless, he provides perspectives on the process of science education that are worth consideration.

Ecker, Ronald L. *Dictionary of Science and Creationism.* Buffalo, NY: Prometheus Books, 1990; 263 pp.
Ecker, a librarian and cotranslator of *The Canterbury Tales* into modern English, provides a dictionary of cross-referenced terms used by scientists and creationists. Ecker's hopes to provide information that will help politicians, educators, and the media counter what he considers a trend toward pseudoscientific perspectives in America's schools. He carefully distinguishes creation science, of which he takes a dim view, from a spectrum of other creationist and evolutionist perspectives he describes under various entries. Definitions are concise and nontechnical.

Eldredge, Niles. *The Monkey Business: A Scientist Looks at Creationism.* New York: Washington Square, 1982; 157 pp.
Eldredge is a paleontologist and a curator of invertebrates at the American Museum of Natural History. He is probably best known for his coauthorship (with Stephen Jay Gould) of the theory of punctuated equilibrium. Eldredge believes "that the beauty and relevance of Genesis 1 are neither threatened nor enhanced by modern science." In fact, he holds that science and religion involve completely different spheres of human experience. Thus, creationism has no place in science. He discusses the nature of science and the predictions of evolutionary theory. He then describes the fossil record, explaining how it satisfies these predictions. After a discussion of biological evolution, he critiques creationist ideas on design, thermodynamics, geology, and paleontology. Creationists, he writes, are "liars" with a "peculiarly myopic view of the natural world." His last chapter deals with the political and social ramifications of the creationist movement, warning readers of its dire effects on America's children.

Hanson, Robert W. (ed.). *Science and Creation.* New York: Macmillan, 1986; 213 pp.
This volume is the outgrowth of a 1982 symposium sponsored by the American Association for the Advancement of Science. The contributors, all educators interested in the issue of creationism in the

public schools, represent a variety of academic disciplines. The first several chapters consider evidence bearing on the issue of earth history. This is followed by presentations of case histories from four states. The final chapters encourage educators to treat controversial issues fairly, to recognize the limitations of science, and to treat the views of students with respect. This is a well-written, carefully edited volume which promotes "a better understanding of the limitations of science and of the nature of religion wherever either is taught."

Harrold, Francis B., and Raymond A. Eve. *Cult Archaeology & Creationism: Understanding Pseudoscientific Beliefs About the Past.* Iowa City: University of Iowa Press, 1987; 163 pp.
This book contains contributions by scholars to a symposium on cult archaeology and creationism held in 1986. It focuses on "unsubstantiated beliefs about the human past." A majority of the chapters report the results of surveys of student beliefs regarding creationism and cult archaeology (e.g., extraterrestrials, the Loch Ness monster, Velikovsky's views). Not only are students influenced by what the authors call "pseudoscience," but researchers show that even teachers are not immune to this influence. One survey, for example, indicated that 11 percent of the science educator respondents believed that the earth is less than 20,000 years old. This is a thought-provoking series of papers with particular import for educators.

Hoodbhoy, Parvez. *Islam and Science.* London: Zed, 1990; 154 pp.
Hoodbhoy, a Pakistani who did graduate work at Massachusetts Institute of Technology, believes that the key to cultural success lies in a commitment to science. He notes that centuries ago, when Islam was strong and rich, it supported the pursuit of knowledge through scientific investigation. At the same time the western world, dominated by the Christian church, was weak and poor, and it devalued science. Today the situation is reversed. Islamic fundamentalism not only rejects Darwinian evolution, he says, but it also supports an earth-centered view of the universe. In fact, in 1982 the president of a Saudi Arabian university published a book defending this view. Hoodbhoy's book provides interesting insights into elements of Islamic fundamentalism. It is clearly written and well referenced.

Hoyle, Fred. *The Origin of the Universe and the Origin of Religion.* Wakefield, RI: Moyer Bell, 1993; 91 pp.

Hoyle was knighted by the British royalty as one of this century's preeminent astronomers and cosmologists. He is also a controversial, iconoclastic figure who is opposed to the concept that life arose spontaneously on earth, and who recently led an ill-fated crusade to discredit the feather impressions of fossil *Archaeopteryx* specimens as fakes. A confirmed non-theist, Hoyle attempts in this book to suggest that religion and mythology, among other things, arose as a result of periodic cometary impacts on the earth. According to Hoyle, these impacts could also have been responsible for melting the permafrost of the ice age, leading to the extinction of the woolly mammoths; for the discovery of smelting by early humans; and for the destruction of the biblical cities of Sodom, Gomorrah, and Jericho. In this book Hoyle demonstrates once again his penchant for defending unpopular and sometimes off-beat ideas.

Hughes, Liz R. (ed.). *Reviews of Creationist Books.* Berkeley, CA: National Center for Science Education, 1992; 147 pp.
This book provides reviews of more than 40 creationist books. The National Center for Science Education published this collection to help teachers, librarians, and parents assess the value of these materials for use in science classes. Books reviewed include those by Michael Denton, Duane Gish, and Henry Morris, and reviewers include Joel Cracraft, G. Brent Dalrymple, Stephen Jay Gould, and Michael Ruse. Reviews point out the limitations of creationist models favoring a young earth, limited evolutionary change, and a worldwide flood, and promote naturalistic approaches to earth history.

Keller, Stephen H. *In the Wake of Chaos: Unpredictable Order in Dynamical Systems.* Chicago, IL: University of Chicago Press, 1993; 190 pp.
Chaos theory has developed through the study of the nonlinear dynamic systems that exhibit deterministic behavior. This theory has been applied to everything from ongoing physiological processes to the origin of the universe and life. In this book, Keller discusses the nature of chaos theory and its philosophical implications. He notes that chaotic systems are very sensitive to initial conditions—that very minor changes in initial values can translate to major differences in final values. Because of this, it is often impossible to predict the future state of a chaotic system with any degree of precision. The best scientists can hope for is to be able to qualitatively predict the types

of changes the system might undergo. In practical terms, Keller notes, we cannot think of the world as deterministic—even imperceptible quantum fluctuations can alter initial conditions in ways that alter outcomes. This book provides an excellent introduction to chaos theory and its implications for how we view science, nature, and the history of the universe.

Montagu, Ashley (ed.). *Science and Creationism.* New York: Oxford University Press, 1984; 416 pp.
This volume, edited by anthropologist Montagu, consists of a collection of twenty essays by scholars in science, history, law, and education. The collection is a response to court battles over the teaching of creationism in public schools, particularly the Arkansas "Scopes II" trial held in 1981. Creationist arguments are presented and countered, and Judge William Overton's decision to strike down the Arkansas law is reprinted. The positive case for evolution is defended. Several authors point out that creationists misunderstand controversies among evolutionary scientists as evidence that the theory of evolution rests on shaky footing. Creationists are also accused of misquoting evolutionists. This book provides a useful collection of arguments used against the 1980's creationist movement.

Newell, Norman D. *Evolution and Creation: Myth or Reality?* New York: Columbia University Press, 1982; 199 pp.
Newell is a retired curator at the American Museum of Natural History, a professor emeritus at Columbia University, member of the National Academy of Sciences, and one of the twentieth century's preeminent paleontologists. This book was written "especially for schoolteachers, young people and their parents, and for all those whose scientific background is not adequate to withstand the high-pressure methods and the misleading arguments posed by the creationists." Newell notes that while many fundamentalist Christians insist on interpreting Genesis in a literal manner, many other Christians, Jews, and Muslims read the Bible "as allegorical and symbolic." After a discussion of patterns of life, Newell examines the beliefs and agendas of creationists and flood geologists and concludes that they misuse the Bible and are self-deceived. Most of the book, however, consists of a description of the fossil record and an elementary presentation of evolutionary biology in an attempt to show the falsity of creationist views.

Plimer, Ian. *Telling Lies for God: Reason vs. Creationism.* Milsons Point, NSW, Australia: Random House, Australia, 1994; 303 pp.
Plimer is a professor of geology at the University of Melbourne. His book constitutes a diatribe against what he sees as the abuses carried out in the name of creation science. He begins asserting that evolution has no relevance to one's faith. He examines the claims of creationists and points out the difference between "science" and "dogma." He expresses outrage toward creation scientists—their selective presentations of data, their misquotations, their abuse of the democratic process, and their notion that both young earth and old earth views are mere theories. Plimer reviews the methods by which scientists determine the age of the earth, examines fraudulent claims by creationists, and critiques what he believes to be the absurdities of creationist claims and logic. In the last chapter, Plimer expresses his amazement that otherwise intelligent people accept the tenets of creationism and notes that creationism has not received support from mainline Christian denominations. The book receives endorsement in a foreword by Peter Hollingsworth, Anglican Archbishop of Brisbane, and in a preface by Robyn Williams, ABC National Science Unit.

Strahler, Arthur N. *Science and Earth History: The Evolution/Creation Controversy.* Buffalo, NY: Prometheus Books, 1987; 552 pp.
Strahler, retired chair of the Geology Department at Columbia University, has provided a *tour de force* for consideration by participants in the evolution/creation controversy. This large volume, containing 54 chapters of well-documented text, two columns to a page, is designed to clarify issues on both sides of the controversy. Strahler first attempts to differentiate science from pseudoscience, then evaluates creation science from the perspectives of cosmology, astronomy, geology and crustal history, origins of landscapes, stratigraphy, and the fossil record. He also discusses evolutionary theory, the emergence of humans, and the origin of life. While not particularly sympathetic toward creation science, he attempts to be fair and gives credit where he thinks it is due. This is a comprehensive and quite readable reference work.

_____. *Understanding Science: An Introduction to Concepts and Issues.* Buffalo, NY: Prometheus Books, 1992; 381 pp.
Strahler sets out to provide readers with a comprehensive philosophy of science. He believes that modern scientific perspectives are neces-

sarily naturalistic, and that the universe and everything it contains appeared without supernatural intervention. He argues that we can never say that anything is true, that ultimately, our scientific propositions are merely probability statements. Strahler conceives of knowledge as belonging to two classes: perceptional and ideational. Science and human history constitute perceptual knowledge, whereas religion and ethics belong to the realm of ideational knowledge. Ideational knowledge is interesting, he admits, but of little practical value. Strahler rejects the notion of the supernatural and believes that ethics and morality arose through evolutionary processes. He is critical of any attempt to wed theistic outlooks to science.

Toumey, Christopher P. *God's Own Scientists: Creationists in a Secular World.* New Brunswick, NJ: Rutgers University Press, 1994; 280 pp.

This is one of the most interesting and unusual books to be published on the creation/evolution controversy in recent years. Toumey, a Catholic anthropologist who is also an evolutionist, met for an extended period of time with a creationist study group in the North Carolina "research triangle." Group members were deeply committed to the conservative creationism espoused by Henry Morris and the Institute for Creation Research. They happily opened their meetings to Toumey. He found that privately they exhibited an amazing diversity of viewpoints on creationism, though publicly they presented a more unified perspective; he also discovered that members were relatively naive about the philosophical and historical roots of their movement. Toumey argues that it is through creationism that many conservative Christians try to make sense of twentieth century social ills—evolution, as they see it, is to blame for these problems. He also contends that Whitcomb and Morris' *The Genesis Flood* (1961) was responsible for convincing fundamentalists to begin using the authority of science to support their claims about earth history. This book is an honest but sympathetic attempt by an outsider to provide a better understanding of the creationist movement.

_____. *Conjuring Science: Scientific Symbols and Cultural Meanings in American Life.* New Brunswick, NJ: Rutgers University Press, 1996; 197 pp.

This book is something of a sequel to Toumey's earlier *God's Own Scientists: Creationists in a Secular World.* In *Conjuring Science*,

Toumey writes about how American culture misappropriates and misunderstands science. He suggests that science today is regarded in much the same way as God in ancient Israelite culture—worshipped and feared but little understood. Just as God was known symbolically to the Israelites, science is known to American culture through symbols that may have little to do with science itself. This, Toumey claims, occurs because of the cultural dissonance created by a historical shift in how science perceives itself—from an earlier outlook informed by Scottish commonsense realism, Baconian empiricism, and Princetonian theology, to one that now subscribes to the secularism, rationalism, naturalism, and autonomy brought about by the European Enlightenment. Toumey posits an anthropological model to account for this dissonance, one that lays blame on television, the American educational process, and postmodern thought. This dissonance, he says, results in the confusing misappropriation of science as, for example, by fundamentalist creation scientists who distort and misuse data, and by tobacco companies that point out that smoking and cancer can be linked only statistically, not causally.

Wilson, David B. (ed.). *Did the Devil Make Darwin Do It? Modern Perspectives on the Creation-Evolution Controversy.* Ames: Iowa State University Press, 1983; 242 pp.

This book is a collection of papers written by professors at Iowa State University on the creation/evolution controversy. Contributors examine the claims of creationists in view of nature and province of science, thermodynamics, history of life, comparative religions, and the meanings of the terms used by creationists and their opponents.

Zetterberg, J. Peter (ed.). *Evolution Versus Creationism: The Public Controversy.* Tucson, AZ: Oryx, 1983; 516 pp.

This volume represents the proceedings of a 1981 conference at the University of Minnesota titled "Evolution and Public Education." It also contains reprints of several articles pertinent to this issue. The papers in the first section describe evolutionary theory and its significance to education. The creationist movement and its main positions are highlighted in the second and third sections, with several reprinted pieces by creationist authors. The fourth section examines ways of responding to the creationist challenge, with an emphasis on discrediting creationist views on such topics as evolutionary transitions, the second law of thermodynamics, the age of the earth, and fossil footprints. The fifth section takes up legal issues and looks specifically at court battles in Tennessee, Minnesota, Louisiana, and Arkansas. The final section examines the "equal time" argument, raises "Six 'Flood' Arguments Creationists Can't Answer," and attempts to elucidate the real issues underlying the creation/evolution controversy.

Chapter 5

PHYSICS AND COSMOLOGY

Physicists examine the characteristics of matter and motion, while cosmologists study the history, structure, and operation of the physical universe. The most fundamental questions asked by physicists and cosmologists are at the very core of the creation/evolution controversy—Where did matter come from, how does matter behave, and what determines its behavior? The books listed below attempt to answer these most basic of questions.

THEIST REFERENCES

Block, David. *Star Watching*. Chicago, IL: Lion, 1988; 160 pp.
Block is an astronomer and devout Christian. This book surveys the universe from the solar system to the most distant realms of space. The author believes that the vastness of the universe is an invitation to search for its creator. He contends that the galaxies were created by an amazing, personal God who searches humans out. He reviews examples of apparent design and posits that the creator planned the universe for the existence of life. Block challenges the notion that modern cosmology makes the notion of God improbable. This is a well-illustrated book in which the author celebrates his belief in God as creator of the cosmos.

Craig, William Lane, and Quentin Smith. *Theism, Atheism and Big Bang Cosmology*. New York: Clarendon Press, 1993; 337 pp.
Craig, a theist, teaches philosophy at the Catholic University of Louvain, while Smith, an atheist, teaches philosophy at Western Michigan University. Their book consists of an extended debate over the origin of the universe. Both authors accept the Big Bang but they differ in their interpretation of this event and its aftermath. Craig

posits that the concept of infinity is a theoretical one, one that cannot be actualized in history. He also states that everything in the universe with a beginning must have a cause, and that the cause of the universe, which is finite, is a creator God. Smith counters by arguing that the Big Bang was a one-time event with no need of a cause. Both authors respond to the cosmological theories of Stephen Hawking. Craig argues that Hawking's views are compatible with theism, while Smith believes they are more in line with atheism.

Davies, Paul. *God and the New Physics.* New York: Simon and Schuster, 1983; 255 pp.
Davies is a professor of theoretical physics at the University of Newcastle-upon-Tyne and a prolific author. In this book he hopes to take the reader toward "a unified description of all creation," which he believes to be within sight of the "new physics." Davies considers all the classical arguments for the traditional God and finds them wanting because they depend on outdated concepts of space, time, and matter. He believes that if we use a conventional God to fill in gaps in knowledge of reality, this conventional God will disappear as science progresses. He considers the notions of mind, soul, self, free will, and determinism, and concludes that life is "an integral part of the cosmic miracle." He is impressed by the fact that the fundamental constants of nature seem to point to some type of cosmic design—the anthropic principle. Davies, however, is not here slipping into conventional theism. Instead, he sees the phenomenon of "mind" as an all-important guiding force in the cosmos.

———. *The Cosmic Blueprint.* New York: Simon & Schuster, 1988; 224 pp.
In this book Davies notes that many contemporary views of cosmology are based in classical Newtonian physics. He believes this approach is too limiting—that the universe is continually self-organizing through principles of physics that have yet to be discovered. He uses living systems to illustrate his point that at higher and higher levels of complexity, emergent qualities become apparent that could not be seen at lower levels of complexity. Davies is impressed by the appearance of design in the universe, and he searches for rational explanations for this design. Ultimately, however, he thinks there may be something going on behind the scenes that gives existence meaning.

Dolphin, Lambert T. *Jesus: Lord of Space and Time.* Green Forest, AR: New Leaf, 1988; 271 pp.
In this admittedly speculative book, Dolphin attempts to link concepts in the Bible with physical reality. For example, he discusses the unorthodox view that the speed of light has decayed exponentially with time, allowing for an old-looking universe to be really young. He also posits that spiritual space and time differ from physical space and time. From this perspective he suggests that at the time of death everyone arrives at eternity at the same time. Throughout the book Dolphin assumes a Platonic dualism which differentiates the illusionary physical world from the real spiritual realm. He also argues that God programmed the world to run "without His constant on-going interference which would show up immediately as violations of the known laws of physics." Dolphin assumes an orderly universe and that science and revelation are in basic harmony.

Drees, Willem B. *Beyond the Big Bang: Quantum Cosmologies and God.* La Salle, IL: Open Court, 1990; 323 pp.
This book is based on a doctoral dissertation by Drees, who is a theoretical physicist. His central concern is the Big Bang theory in relation to theology. After examining the theory of the Big Bang and concepts of creation, he suggests that it is unwise to make theological arguments on the basis of Big Bang cosmology; likewise, it is inappropriate to argue against contemporary cosmologies from a creationist position. Drees believes that the Big Bang theory is religiously neutral. He discusses the anthropic principle and posits that the appearance of design neither supports nor refutes theistic notions. He also points out that scientific theories, like the Big Bang, are not proven facts but only tentative proposals. Drees believes that although theology is ambiguous, scientific perspectives regarding the universe are limited. Despite these problems, he calls for the construction of a carefully reasoned consonance between cosmology and theology.

Heeren, Fred. *Show Me God: What the Message from Space Is Telling Us About God. Wonders That Witness, Volume 1.* Wheeling, IL: Searchlight Publications, Daystar Productions, 1995; 337 pp.
This book is an attempt by a Christian apologist to provide a factual basis to convince skeptics, Christian and otherwise, that there is reason to believe in the God of the Bible. Heeren does this by exploring

fundamental cosmological questions common to both science and religion. Through interviews with contemporary cosmologists such as Stephen Hawking, Robert Jastrow, Arno Penzias, and Robert Wilson, Heeren brings a sense of credibility to his arguments. In the 12 chapters he discusses logical support for the existence of a creator God, alternatives to the Big Bang, various creationist positions, chance and design, the anthropic principle, and the Gospel in relation to cosmology. He is critical of attempts to understand creation exclusively through the Bible without recourse to the creation itself. He attempts to capture the interest of readers in a number of ways. For example, scattered throughout the book are pieces of an imaginary conversation with a Christian book publisher who is more interested in sales than in truth; Chapter 3 consists of a science fictional account about messages from extraterrestrial intelligences; and a section toward the end of the book contains descriptions of 50 important scientists who believed in God.

Houghton, John. *Does God Play Dice? A Look at the Story of the Universe.* Grand Rapids, MI: Zondervan, 1989; 160 pp.
Houghton, director general of the Meteorological Office of England and former head of the Department of Atmospheric Physics at Oxford, attempts to integrate two important aspects of his life—his science and his Christianity. He discusses a variety of topics in the physical sciences, such as the Big Bang and particle theory, and relates these to the spiritual realm. Houghton believes that God's activity can take place in a dimension beyond that of time and space, making it difficult for humans to perceive this activity. He refers frequently to the writings of C. S. Lewis and Donald MacKay.

Humphreys, D. Russell. *Starlight and Time: Solving the Puzzle of Distant Starlight in a Young Universe.* Colorado Springs, CO: Master Books, 1994; 133 pp.
Humphreys is a physicist. He believes that the entire universe is only a few thousand years old and provides what he believes to be the evidence for this view. He utilizes the general theory of relativity to support his contention that stars millions of light years away were created on the fourth day of creation and that light from these stars reached the earth at that time. He points out weaknesses in the Big Bang theory and in the cosmologies of earlier young earth creationists. According to Humphreys, water was the first matter in the uni-

verse to be created. Appendices contain the text of two papers delivered by Humphreys at the 1994 International Conference on Creationism. The first of these papers provides a biblical basis for his views and the second develops the underpinning of his young earth relativistic cosmology.

Miller, James B., and Kenneth E. McCall (eds.). *The Church and Contemporary Cosmology.* Pittsburgh, PA: Carnegie Mellon University Press, 1990; 400 pp.
This book was produced by the Task Force on Theology and Cosmology of the General Assembly of the Presbyterian Church, USA. Its purpose is to explore changes in cosmology in relation to the Bible and theology. Contributors include noted scholars such as Langdon Gilkey, Ian Barbour, and 17 other writers. Papers consider the ancient Israelite cosmology, the New Testament concept of heaven, recent developments in the physical and biological sciences, and contemporary interactions between science and theology. Barbour provides a summary chapter in which he describes four ways of relating science and religion: conflict, independence, dialogue, and integration. He also discusses three significant scientific and theological periods: Medieval, Newtonian, and the twentieth century. This book provides a wealth of information on contemporary issues of sciences and religion.

Mulfinger, George (ed.). *Design and Origins in Astronomy.* Norcross, GA: Creation Research Society Books, 1983; 152 pp.
Mulfinger is a frequent contributor to the creationist literature on astronomy, and this book contains an assemblage of creationist views on this topic. The first section examines the universe from the teleological perspective that the universe was carefully designed for the purpose of inhabitation. The second section addresses the issue of the origin of the universe and questions the Big Bang theory and its underlying notion of the expanding universe and the meaning of the redshift. The third section provides a description of the solar system and questions the assumption that the sun's energy arises out of nuclear reactions, as well as the assertion that the sun's life is of a finite duration. A concluding section asserts that the Bible is not in conflict with astronomy, but that it anticipated in a remarkable way the findings of modern astronomy.

Parker, Barry. *Creation: The Story of the Creation and Evolution of the Universe.* New York: Plenum, 1988; 295 pp.
Parker, a physicist at Utah State University, reviews the history of the development of the contemporary scientific cosmology. He highlights important personalities and discoveries in physics as they relate to the development of modern cosmology. He explains the difficulties of trying to merge the concepts of theoretical physics with the results of experimental physics. Parker believes that scientists must do their work without reference to God, but even if they someday develop a complete scientific explanation for the origin of the universe, they still will not have explained how the basic laws of nature came into existence. He believes that because of this there will always be a need for the concept of God. This is a well-written, nicely illustrated book for the general reader.

Ross, Hugh. *The Creator and the Cosmos: How the Greatest Scientific Discoveries of the Century Reveal God.* Colorado Springs, CO: NavPress, 1993; 155 pp.
Ross, who has a Ph.D. in astronomy, directs a ministry called "Reasons to Believe" devoted to searching for harmonies between the Bible and nature. This search for harmony is the purpose of this book. He purports to have made a calculation that the Bible is 10^{58} times more reliable than the second law of thermodynamics. He believes that Genesis 1 provides a scientifically accurate summary of the origin of the earth from an astrophysical and geophysical standpoint. In his view Christians must choose between naturalistic explanations for how things happened and acts of God that cannot be scientifically described. He seems to accept the notion, however, that over time the forms and functions of organisms have changed to some degree as a result of natural processes. He suggests that "No society has seen as much proof for God as ours."

Van den Beukel, Anthony. *The Physicists and God: The New Priests of Religion?* Translated from Dutch by John Bowden. North Andover, MA: Genesis, 1995; 182 pp.
The author is a Dutch scientist, and this book was originally published in The Netherlands in 1990. The author shares with his readers a variety of personal musings over issues related to science and religion. Of particular concern is whether or not objective reality exists. According to Van den Beukel, the observed and the observer are one

and the same. Van den Beukel examines the views of Isaac Newton, Blaise Pascal, Albert Einstein, and Stephen Hawking in relation to belief in the existence of God. He rejects the notion that one can find God through science or that the existence of God can be proved. He does believe that the existence of God and the impact of God on peoples lives "is beyond any reasonable doubt." Toward the end of the book Van den Beukel looks at the interactions between science and faith but concludes that there "is no such thing as Christian physics." He also considers the social responsibilities of scientists, although he suggests that it is impossible to know how one's work will be used by others.

Whitcomb, John C. *The Bible and Astronomy.* Winona Lake, IN: BMH Books, 1984; 32 pp.
Whitcomb, a professor of Old Testament at Grace Theological Seminary, believes that the primary purpose and message of the universe is theological, not scientific. He believes that the entire universe was created in six literal days. He argues that the creation of the sun, moon, and stars on the fourth day of creation completely precludes any theory of an evolutionary origin for the universe. He examines the Bible for evidence of intelligent beings on other planets and concludes that no such evidence can be found. In fact, according to Whitcomb, the Bible seems to suggest that no such beings exist. Christians, he writes, should not attempt to understand the biblical stories of Hezekiah's sundial and Joshua's long day in scientific terms because they are beyond the realm of scientific explanation—they were completely miraculous.

Wilkinson, David. *God, the Big Bang, and Stephen Hawking.* Second edition. Crowborough, England: Monarch, 1996; 176 pp.
Wilkinson, a theoretical astrophysicist, fellow of the Royal Astronomical Society, and Methodist minister, provides here a response to Stephen Hawking's book *A Brief History of Time* from the perspective of the Christian faith. He first summarizes contemporary views on cosmology and physics, with a focus on issues like the origin and evolution of the universe, quantum mechanics, relativity, the anthropic principle, and chaos theory. He then evaluates these notions in view of his faith as a Christian. Wilkinson believes that the Christian faith has something to give and something to receive in its relation to science and cosmology. He notes that Hawking's blurring of

temporal and spatial time, his "no-boundary condition" which postulates no beginning of time for the universe, and his thoughts on a "theory of everything" are controversial and should not be taken as the last word. He also criticizes some of Hawking's views on religion and his assumption that religion and science are bound to clash. In an appendix, Wilkinson takes issue with young earth creationism and fosters the view that Genesis 1 should be interpreted theologically, not scientifically. The book is written for someone who has little background in science or religion and contains some useful illustrations.

NONTHEIST REFERENCES

Barrow, John D. *The World Within the World.* New York: Oxford University Press, 1988; 398 pp.

Barrow, an astronomer at the University of Sussex, examines several views of science, including empiricism, operationalism, idealism, and realism. He reviews the history of science and suggests that although the biblical tradition did not lead directory to science, its demythologization of nature and its view of a lawful God provided a supportive framework for the development of scientific thought. Barrow attempts to link the concepts of elementary particles and of cosmology and provides an explanation for why the laws of nature are mathematical. He looks at ways in which humans select particular phenomena as they seek to understand the order of the universe. He rejects the notion that the anthropic principle argues for the existence of God but notes that this principle is not incompatible with such a view. Barrow points out that any "theory of everything" is ultimately untestable, given that we have no universe other than our own to test it against.

———. *Theories of Everything: The Quest for Ultimate Explanation.* New York: Oxford University Press, 1991; 223 pp.

Here Barrow examines the quest for a "Theory of Everything" (TOE) in relation to epistemology, philosophy, and religion. He begins by pointing out that the desire to unify knowledge is religiously motivated. He explores the laws of nature as the most important components of any TOE, then considers the significance of initial conditions. He raises questions about the so-called constants of nature: Why do they have the particular values they have and are they really constants? He discusses chaos and chance in relation to the predictability from a TOE. He looks at the notion of organization in nature, particularly among living things, and notes that a TOE will not help us understand the origin of life or consciousness. Finally, he examines the role of mathematics in the development of a TOE and suggests that some aspects of human experience are beyond the explanatory capabilities of any TOE or mathematical algorithm.

_____, and Joseph Silk. *The Left Hand of Creation: The Origin and Evolution of the Expanding Universe.* New York: Oxford University Press, 1993; 262 pp.
In this book, astronomers Barrow and Silk provide an interesting introduction to cosmology for the educated non-specialist. First, the authors discuss the antiquity of the universe and the Big Bang, including the concept of "infinite density." This is followed by a description of the four fundamental forces—weak, strong, electromagnetic, and gravitational—and their relation to the beginnings of the universe. The origin and structure of galaxies is covered next, followed by a discussion of the expanding universe, time, and the anthropic principle. Although the authors are critical of the argument from design for the existence of a creator, they note that the universe seems "tailor-made for life." The book is helpfully illustrated and contains a useful glossary.

_____, and Frank J. Tipler. *The Anthropic Cosmological Principle.* New York: Oxford University Press, 1986; 706 pp.
Barrow and Tipler, a mathematical physicist at Tulane University, examine in exhaustive detail the nature and implications of the Anthropic Principle. The so-called weak anthropic principle states that the universe is structured in ways compatible with human origin and existence. Put another way, if the universe were structured differently than it is, we would not be here. The strong anthropic principle states that a universe supportive of human life is the only possible universe. Both the weak and strong versions of this principle, the authors note, could be interpreted as evidence for the necessity of a designer, but they prefer a secular interpretation based on an evaluation of quantum mechanics. The authors use the anthropic principle to argue against the existence of intelligent extraterrestrial life. This is a massive, well-referenced, and closely argued book.

Chaisson, Eric. *The Life Era, Cosmic Selection and Conscious Evolution.* New York: The Atlantic Monthly Press, 1987; 259 pp.
Chaisson, a research physicist at MIT and a professor at Harvard and Wellesley, suggests that just as the Darwinian revolution freed us from the belief that we are substantially different from other organisms, so our understanding of cosmic evolution will help us see that our bodies are of the same matter as the distant stars and galaxies. Chaisson believes that the concept of "spontaneous creation" is nonsensical, that fundamentalists should not gain legislative power, and that discussing the Bible as literal history in the classroom is a mistake. He posits that we must begin now to help guide the evolutionary process if we are to have a future. He promotes the notion of a worldwide society that rejects authoritarianism, communism, and totalitarianism and promotes the values of knowledge and compassion.

Dauber, Philip M., and Richard A. Mullet. *The Three Big Bangs: Comet Crashes, Exploding Stars, and the Creation of the Universe.* Reading, MA: Addison-Wesley, 1996; 207 pp.
Dauber and Mullet are physicists, and this book is intended as a supplementary text for courses in physics and astronomy. A central theme is cosmic violence and its role in creating conditions suitable to life. Part I examines collisions of comets and asteroids with planets. Here the authors discuss the asteroid impact that may have brought an end to the Cretaceous Period and the dinosaur dynasty. Part II discusses exploding stars and the generation of elements for the development of life on Planet Earth. Part III looks at the history of the universe. The authors acknowledge the possibility that a divine being may have been responsible for the creation of the universe and the molecules of life. The 23 chapters of this book are written in a popular, easy-to-read style.

Gribbin, John. *In the Beginning: The Birth of the Living Universe.* New York: Little, Brown, 1993; 288 pp.
Gribbin, an astrophysicist and science writer, here makes a case for the view that the universe is "alive—literally, not metaphorically." He first examines the implications of Hubble's law and the special and general theories of relativity for notions of how the universe began at the time of the Big Bang and the importance of dark matter in keeping the universe from flying apart. He follows this by considering life and its evolution, especially the role played by natural selec-

tion acting on mutant forms of DNA. He then examines the nature of the universe, particularly in light of the "Goldilocks effect," that everything seems optimally designed for the existence of life. Finally, Gribbin addresses the question of why our universe appears so well designed for life. He posits that ours is only one of an infinite number of universes some of which, like our own, would be suitable for the development and support of life.

Hawking, Stephen W. *A Brief History of Time: From the Big Bang to Black Holes.* New York: Bantam, 1988; 198 pp.
Considered to be the most brilliant theoretical physicist since Albert Einstein, Hawking is a professor of mathematics at Cambridge University. He wrote this book as an attempt to popularize concepts of space and time for non-specialists. He says that he was told that each equation he "included in the book would halve the sales," so he only included one, Einstein's $E=mc^2$. He begins by showing how human conceptions regarding the universe changed from the time of Aristotle to today. He then addresses questions about the beginning and end of time, the extent of the universe, and the discoveries and theories that impinge on these questions. He explains Einstein's general theory of relativity, quantum mechanics, and efforts to develop a "unified theory of everything," and he applies these ideas to what we know about the cosmos. He concludes by suggesting that if we do develop a unified theory, we will able to ask why the universe exists, and the answer would allow us to "know the mind of God."

Lerner, Eric J. *The Big Bang Never Happened.* New York: Random House, 1991; 466 pp.
In this book written for a popular audience, Lerner argues against the Big Bang cosmology. He begins by addressing some of the scientific problems with the Big Bang model, then launches into a discussion of the impact of various cosmologies on human societies. He claims that cosmological views influence economic practices, social norms, standards of living, the occurrence of slavery, and other societal factors. Moreover, he suggests that the standard of living in industrialized countries has declined since 1973, largely due to the influence of the finite Big Bang cosmology. By contrast, he claims that infinite cosmologies lead to cultural and economic progress. Lerner supports an infinite cosmology based in plasma physics. Many of the features of the universe, such as spiral galaxies and quasars, result from cur-

rent densities, he says. This cosmological perspective views the universe as existing in infinite time and space, although Lerner accepts the theories of cosmic, biological, and societal evolution.

Smoot, George, and Keay Davidson. *Wrinkles in Time.* New York: Morrow, 1993; 331 pp.
Smoot is the team leader of the Cosmic Background Explorer (COBE) satellite project to examine variation in the microwave background radiation of the universe. This book reports on Smoot's experiments on this topic using high altitude ballons, high-flying planes, and satellites. It highlights the discovery by his COBE team of variability in microwave background radiation, a major prediction of the Big Bang theory. The book includes a list of the corporate sponsors of COBE, as well as the names of the large number of people who participated in the project. The authors note the artistry and order apparent in the universe but suggest that if there is a God, he remains hidden behind the reality of the Big Bang.

Stewart, Ian. *Does God Play Dice? The Mathematics of Chaos.* Cambridge, MA: Blackwell, 1989; 317 pp.
This book really has nothing to say about God, but it has a lot to say about chaos. Chaos, in today's scientific lexicon, refers to the complicated, stochastic behaviors that arise in deterministic systems. Chaos theory has impacted virtually every science, from physics to biology to cosmology. Steward provides an interesting, intuition-based introduction to chaos. He begins by exploring the history of ideas leading to chaos theory, from Newton's celestial mechanics, through recent notions of chance variation around means, to the computer revolution, which simplified the calculations of chaos theory and allowed for its visualization. He then shows how chaos theory can be applied to a diverse array of systems from dripping faucets to tumbling moons. In the final chapter Steward discusses the relationship of the concept of chance to to the concept of chaos. This is a well-written book intended for anyone interested in how contemporary science explains the origin of complex systems.

Swimme, Brian, and Thomas Berry. *The Universe Story: From the Primordial Flaring Forth to the Ecozoic Era—A Celebration of the Unfolding of the Cosmos.* San Francisco, CA: Harper, 1992; 305 pp.
Swimme is a specialist in mathematical cosmology and Berry is a

historian of cultures. This book is an attempt to create an epic tale of the history of the universe, from the Big Bang to the development of human culture and civilization. This "Great Story" is told without quotations, scientific equations, or footnotes, yet is true to the latest scientific and historical findings. The authors believe that there are three "governing tendencies" of the universe: differentiation, inner spontaneity, and intimate bonding. The human, they posit, is the form by which "the universe reflects on and celebrates itself." They believe the story of the universe is a necessary context for professionals in medicine, law, religion, economics, and education. Swimme and Berry suggest that we are entering a new biological period they call the "Ecozoic Era," a time they hope will be used to enhance the relationship between humans and the rest of the universe.

Weinberg, Steven. *The First Three Minutes: A Modern View of the Origin of the Universe.* Updated edition. New York: Basic Books, 1988; 198 pp.

When this book was first published in 1977, the general public first became aware that scientists thought they could now determine what happened during the initial few minutes after the Big Bang. A chapter on cosmology since 1976 has been added to the new edition. Weinberg first explains how the work of Edwin Hubble, who discovered redshifts, and of Arno Penzias and Robert Wilson, who discovered cosmic background radiation, led to the expanding universe theory. He then develops his own story about the early history of the cosmos, a story that melds the expanding universe concept with inferences from particle physics. Weinberg does not believe that humans are any more than a "farcical outcome of a chain of accidents" and that from the perspective of science the existence of the universe "seems pointless." Despite the pessimism, this is a well-written book by a Nobel Prize-winning physicist.

Chapter 6

EARTH SCIENCE

One of the most significant paradigm shifts in recent history has been the acceptance by the majority of the scientific community of plate tectonic theory—the view that huge crustal plates move around on the earth's surface, colliding, slipping past, or riding over and under one another. This theory purports to explain everything from the eruption of Mount St. Helens to the preponderance of marsupials in Australia. In the face of this powerful explanatory tool a persistent group of creationists, some with Ph.D.'s in geology from prestigious universities, continue to make the radical claim that Noah's flood was responsible for most of the surface features of the earth. The following references examine evidence concerning the earth and its contested history.

THEIST REFERENCES

Ackerman, Paul D. *It's a Young World After All.* Grand Rapids, MI: Baker, 1986; 131 pp.

Ackerman, a member of the Psychology Department at Wichita State University and president of the Creation Social Science and Humanities Society, concludes from his discussion of geological, geophysical, and astrophysical phenomena that the earth is only a few thousand years old. He arrives at this conclusion through loose discussions of moon dust accumulation, polonium halos, meteorite densities, plate tectonics, sedimentology, and a variety of other topics bearing on geochronology. The author's lack of expertise in the subject matter he purports to discuss will be apparent to knowledgeable readers.

Austin, Stephen A. (ed.). *Grand Canyon: Monument to Catastrophe.* Santee, CA: Institute for Creation Research, 1994; 284 pp.

Austin is a geologist on staff at the Institute for Creation Research. This book introduces various natural features of the Grand Canyon National Park from a creationist point of view. Chapters report on atmospheric studies, archaeology, fossils, and the living organisms of the region. Primary emphasis, however, is on the geology of the canyon in a set of chapters by Austin. He believes that the limestones, sandstones, and shales of the canyon were deposited rapidly by the Genesis flood. Shrinkage cracks, animal trackways, and erosional surfaces between layers are all interpreted from a catastrophist perspective. Austin proposes that the canyon itself was cut after the conclusion of the flood by the catastrophic drainage of three post-flood lakes formed on top of the Colorado Plateau. He offers a model to explain the sequence of events leading to this drainage. He points out some of the inconsistencies in the radiometric dates assigned to Grand Canyon rocks and presents a creationist geologic timetable for the canyon.

DeYoung, Donald B. *Weather and the Bible: 100 Questions and Answers.* Grand Rapids, MI: Baker, 1992; 162 pp.
DeYoung, a physicist at Grace College and editor of the *Creation Research Society Quarterly,* writes here of weather past, present, and future. He begins by examining the fundamental principles of meteorology including atmospheric chemistry, pressure, and temperature, as well as large-scale processes such as the water cycle, jet streams, and the development and function of major weather systems. He also considers more localized weather processes. Where possible, he relates his topic to the Biblical record, such as in a discussion of storms on the Sea of Galilee and an evaluation of climate during the time of Jesus. A chapter titled "Past Weather" outlines DeYoung's creationist perspectives on such topics as preflood weather, glaciation, and reasons for the extinction of dinosaurs. The book is written for the general public in a question-and-answer format.

Dillow, Joseph C. *The Waters Above: Earth's Pre-Flood Vapor Canopy.* Chicago, IL: Moody, 1981; 426 pp.
Scientific creationists have long postulated the existence of an antediluvian vapor canopy enclosing the earth. This canopy, they posit, would have moderated the climate and provided a source for much of the water in Noah's flood. Dillow's book provides a model for this canopy. Dillow believes that the Bible can be used as a modern science textbook. He uses several verses in Genesis to support the no-

tion of a pre-flood canopy and finds similar support for his view in ancient mythology. Dillow's model postulates a canopy with 40 feet of precipitable water. This calculates out to a half inch of rain per hour falling continuously over the biblical 40 days and nights of the flood. Dillow makes several predictions based on his model including (1) an ancient greenhouse effect, (2) a sudden drop in polar temperatures, (3) a high pre-flood atmospheric pressure, and (4) a global flood. He critiques a variety of canopy models by other authors and also points out a number of difficulties with his own model. The most important problem, he says, is heat load in the lower atmosphere resulting from condensation and precipitation. Dillow also considers the problem of seeing stars through the thick vapor canopy but calculates that 255 stars would be bright enough to see at one time. A significant portion of the book addresses the implications of his model for mammoths and other animals found frozen in Siberia.

Friedrich, Orval. *The Great Ice Sheet and Early Vikings in Mid-America.* Elma, IA: Self-published, 1993; 122 pp.

Friedrich is an ordained minister who worked for 12 years as a soil scientist before his seminary training. He hypothesizes that the Pleistocene ice sheet was much larger than generally believed, extending as far south as Louisville, Memphis, San Antonio, and Denver. He calls this extended version of the ice expanse the Great Ice Sheet. He believes that parts of the Great Ice Sheet were still present in the American midwest until 1000 A.D. Between 1000 and 1400 A.D., water from the melted Great Ice Sheet formed what he calls the Melt Water Sea, which extended to the "eastern Dakotas, western Minnesota, [and] northern Iowa." He believes that Viking ships plied the waters of this sea and were anchored to rocks with mooring holes bored through them, rocks still found in Minnesota, Iowa, and North Dakota. In short, Friedrich proposes that the end of the ice age occurred much more recently than generally believed by geologists.

Gentry, Robert V. *Creation's Tiny Mystery.* Second edition. Knoxville, TN: Earth Science Associates, 1988; 348 pp.

This book is the story of Gentry's personal, sometimes painful, odyssey to find scientific evidence that the earth was created in a recent six-day creation. The story centers on pleochroic halos, concentric colored rings that surround radioactive particles in minerals such as mica. Gentry has found some unusual halos that he claims provide evidence that all the rocks of the earth were created instantaneously a few thousand years ago. His interpretation has been questioned by creationists and non-creationists alike, and reasonable alternative hypotheses to explain his discoveries have been suggested, but no one questions Gentry's skill at the bench. For eight years he worked at Oak Ridge National Laboratories and his research results have been published in peer-reviewed journals, including *Science.*

Morris, Henry M. *Science, Scripture and the Young Earth: An Answer to Current Arguments Against the Biblical Doctrine of Recent Creation.* San Diego, CA: Creation-Life, 1983; 34 pp.
This small book was written in response to Davis Young's *Christianity and the Age of the Earth* (1982). Morris begins by critiquing Young's affinity for uniformitarian thinking and for not interpreting Genesis literally. He then discusses some of the scientific evidence used by Young to support an old earth view. For example, Morris claims that radiometric dating is invalid because the initial conditions of rock formation cannot be known, that rock systems are not necessarily closed systems, and that radioactive isotope decay rates could have varied through time. He also suggests that the occurrence of coral reefs, evaporite deposits, and lake varves—all of which seem to indicate the passage of considerable time—are best explained by rapid catastrophism. Morris quotes liberally from the scientific press.

_____, and John D. Morris. *Science, Scripture, and the Young Earth.* Revised edition. El Cajon, CA: Institute for Creation Research, 1989; 95 pp.
John Morris has a doctorate in geological engineering. In this book, he and his father respond to attacks against young earth creationism by evangelical Christians who accept evolution and old earth geology. They point out that many of these attacks are made against *The Genesis Flood,* which is now out of date, and that the attackers have failed to examine more recent creationist research supportive of a young earth. They note that modern geology is more and more prone to accept catastrophic interpretations for the rock record, interpretations more in line with those of flood geologists. They discuss problems with radiometric dating and suggest that those who believe in an old earth pick and choose radiometric dates that are consistent with their preconceived notions while ignoring other dates. Evidences for a young earth are discussed, particularly in relation to the decay of the

earth's magnetic field. Creationist models of this decay suggest that the earth is no more than ten thousand years old. The Morrises believe that flood geology provides the best explanatory model for the fossil record. They decry any attempt by Christians to accommodate their views with those of naturalistic scientists and posit that the most dependable truth about the history of the earth comes through biblical revelation.

Morton, Glenn R. *The Geology of the Flood.* Dallas, TX: DMD Publishing, 1987; 176 pp.
Morton is a conservative Christian geologist who favors a literalistic interpretation of the Bible, but he also takes the geologic data seriously. He critiques a variety of attempts to find concordance between Genesis and geology, then offers his own model based on a supposed increase in "permittivity"—the dielectric constant of free space—during and after Noah's flood. According to Morton, this increase affected such disparate factors as radioisotope age, the fossil record, and astrophysics. Most significant, it caused the diameter of the earth to expand by a factor of two. He suggests that the Genesis genealogies do not represent a continuous patriarchal lineage, and he dates the creation to sometime between 125,000 and 14,000,000 years ago.

Oard, Michael. *An Ice Age Caused by the Genesis Flood.* El Cajon, CA: Institute for Creation Research, 1990; 243 pp.
Oard is a meteorologist who has worked for more than 15 years as a forecaster with the National Weather Service. He has also written extensively for creationist publications. He posits that an ice age could not form under the conditions generally postulated by uniformitarian geologists. Instead, "sustained cooling and heavy snow" would be necessary. He argues that these conditions would most likely occur following a major, destabilizing catastrophe. The Genesis flood is a reasonable candidate for such a catastrophe, he writes, because after this event the atmosphere would have contained large quantities of volcanic dust, had heavy clouds, and exhibited a higher albedo, all favoring heavy snow accumulations. Oard rejects the theory of multiple ice advances and provides calculations to suggest that the continental ice sheets could have developed—then disappeared—within one thousand years. He also believes that the mammoths went extinct because they became trapped by the growing glaciers.

Vardiman, Larry. *The Age of the Earth's Atmosphere: A Study of the Helium Flux Through the Atmosphere.* El Cajon, CA: Institute for Creation Research, 1990; 32 pp.
Vardiman is an atmospheric physicist and a young earth creationist. In this booklet he notes that the noble gases "are of most interest in questions regarding the age of the earth." Vardiman focuses on the flux of one of these gases in the atmosphere, helium. He notes that there are two possible sources for atmospheric helium: leakage of helium from the earth's crust and primordial helium present from the time of creation. He argues that helium should enter the atmosphere from the crust at a higher rate than thermal escape of helium from the atmosphere. The helium content of the atmosphere is significantly lower than expected on the basis of an old earth model. Vardiman contends that none of the standard scientific explanations for this apparent anomaly have been adequately quantified or tested. He provides a young earth creationist model of helium flux which assumes that most of the helium in the atmosphere was created with the world fewer than ten thousand years ago.

_____. *Sea-Floor Sediment and the Age of the Earth.* El Cajon, CA: Institute for Creation Research, 1996; 94 pp.
In this book Vardiman seeks to interpret sea floor sediments in terms of his young earth, flood geology perspectives. He posits that the Upper Cretaceous sediments were deposited at the conclusion of the flood. His chief concern is with the deposition of what he considers to be post-flood, Tertiary ocean sediments. He interprets oxygen isotope ratios from foraminiferan shells as indicating a dramatic drop in the temperature of the sea water after the flood; this, in turn, resulted in rapid oceanic circulation and upwelling which stimulated post-flood biological productivity and high levels of sedimentation. It was these high levels of sedimentation, he believes, that produced the Tertiary deposits recovered by sea floor drilling activity. According to his model, however, sedimentation rates have declined exponentially over time until the present day. In addition to his model, Vardiman provides an overview of the sea floor drilling program and data on the nature, thickness, and distribution of sea floor sediments.

Wonderly, Daniel E. *Neglect of Geologic Data: Sedimentary Strata Compared with Young-Earth Creationist Writings.* Hatfield, PA: Interdisciplinary Biblical Research Institute, 1987; 130 pp.

Wonderly, a conservative Christian, provides an enormous amount of evidence to suggest that many of his fellow believers are wrong to insist that the earth and life are only a few thousand years old. Not only does he attempt to show that young earth creationists misinterpret the geological evidence, but he also provides an in-depth discussion of four geological processes that point to a very old earth: (1) deposition and subsequent alteration of strata, (2) the great thicknesses of sedimentary rocks, (3) lithification of successive layers of dissimilar strata and their enclosed fossils, and (4) deposition of evaporite (salt) layers. Wonderly patiently explains why he believes Noah's flood could not have been responsible for these deposits. He then suggests reasons for why many creationists continue to support young earth concepts, including their isolation from mainline scientists, their poor understanding of geologic processes, their disregard of data unsupportive of their views, their distrust of contemporary science, and their narrow interpretation of the Bible.

Young, Davis A. *Christianity and the Age of the Earth.* Grand Rapids, MI: Zondervan, 1982; 188 pp.

Young, a professor of geology at Calvin College, is concerned that Christians are being led astray by the unfounded arguments of young earth creationists. He examines three issues bearing on the age of the earth question in the Christian community. First, he traces the history of Christian views related to the age of the earth. Most Christians had accommodated to an old earth view by the end of the nineteenth century, but beginning in the 1920's a reactionary movement set in, leading to publication of Whitcomb and Morris' *The Genesis Flood* (1961) and a resurgence of young earth creationism. Young believes this reversion to pre-nineteenth century views is mistaken on both scientific and biblical grounds. In the second part of his book Young explains why he believes young earth creationism and flood geology are on the wrong track. He examines a variety of evidences from stratigraphy, sedimentology, and geochemistry that seem to preclude a young earth interpretation. Much of the evidence used to support flood geology, he asserts, is best interpreted by a long age model. The third section of the book focuses on philosophical issues. Young believes that nature and scripture have the same author and that when they are both rightly interpreted there will be no conflict. The conflict over the age of the earth in the Christian community, he believes, is due to faulty biblical interpretation, not faulty science.

_____. *The Biblical Flood: A Case Study of the Church's Response to Extrabiblical Evidence.* Grand Rapids, MI: Eerdmans, 1995; 327 pp.
Young examines, in chronological fashion, the many ways in which Christians have interpreted scientific data in relation to the flood story in Genesis. He is critical of many recent evangelical writers who, in their zeal to support the faith of their readers, have overlooked or misinterpreted a wealth of scientific data. He suggests that flood geologists have often appealed to long-discredited theories to help explain the data they do choose to examine; moreover, they have often assumed miraculous interventions to explain away the most vexing problems raised by the evidence. According to Young, if Christians hope to be effective witnesses in today's increasingly knowledgeable culture, they must speak truthfully and knowledgebly about the sciences, including geology and paleontology, as they relate to the Christian faith.

NONTHEIST REFERENCES

Carey, S. W. *Theories of the Earth and Universe: A History of Dogma in the Earth Sciences.* Stanford, CA: Stanford University Press, 1988; 413 pp.
Carey is one of the last and most vocal opponents of the theory of plate tectonics. This book is a polemic against this theory and substitutes instead an expanding earth model to account for geologic processes. Carey points to a variety of data that seemingly contradict plate tectonic theory. A major difference between his model and plate tectonics is the ultimate disposition of rocks produced by oceanic rift zones and other source areas. Plate tectonic theory posits that this rock is eventually recycled downward into the earth at subduction zones. Carey's model, by contrast, has these rocks pushed upward and outward leading to global expansion. Carey also takes issue with the Big Bang origin of the universe, believing instead in an ill-defined "null universe" model. In this view, a continual addition of energy and mass in equal and canceling ("null") increments leads to the expanding universe. Carey's book provides an interesting and radical critique of contemporary scientific models from an informed perspective.

Chapman, Clark R., and David Morrison. *Cosmic Catastrophes.* New York: Plenum, 1989; 302 pp.

This book is written to explain the history of catastrophism to nonspecialists. The authors demonstrate that, while geology had once been viewed from a catastrophist perspective, the writings of James Hutton, Charles Lyell, and others shifted the emphasis of geology toward a decidedly uniformitarian perspective. Uniformitarianism reigned as the supreme dogma of geology until the 1950's, when Eugene Shoemaker and others correctly identified the presence of meteorite craters on Earth, the moon, and Mars. The authors discuss the Alvarez hypothesis, which explains the presence of an iridium layer at the top of the Cretaceous strata and the demise of the dinosaurs as having resulted from a large meteorite impact. The possibility that impacts recur on a periodic basis is also discussed. The sensational, catastrophist views of Immanuel Velikovsky are referred to as catastrophism gone wild, and creationism is discussed as fringe catastrophism.

Dalrymple, G. Brent. *The Age of the Earth.* Stanford, CA: Stanford University Press, 1991; 474 pp.
Dalrymple is a research geologist with the U.S. Geological Survey. He first encountered "the creationists' peculiar arguments about nature" in 1975, when Henry Morris and Duane Gish presented the case for young earth creationism to a group of professional geologists assembled at an evening seminar in Menlo Park, California. Later he participated in legal proceedings over the creation/evolution issue in California, Louisiana, and Arkansas, and he has written and spoken extensively on the topic of the age of the earth in this context. While the body of this book does not directly address the arguments used by creation scientists for a young earth—arguments he considers "absurd"—the impetus for this book grew out of Dalrymple's involvement with the creation/evolution controversy. Specifically, his purpose "is to explain how scientists have deduced the age of the Earth." He first examines contemporary theories about the history of the universe and recounts the series of attempts to determine the age of our planet before the development of radiometric dating techniques. He then discusses the phenomenon of radiometric isotope decay processes and shows how scientists have harnessed these processes to date the earth and the solar system. He also explains how estimates for the ages of the Milky Way and the entire universe have been determined. A concluding chapter summarizes information from the previous chapters, suggests where age-of-the-earth research may profitably

lead in the future, and posits that it is unlikely that future age estimates will differ dramatically from those of today. Dalrymple wrote the book "for people with some modest background in science" and tried to minimize the use of mathematics.

Dott, Robert H., Jr., and Donald R. Prothero. *Evolution of the Earth.* Fifth edition. New York: McGraw-Hill, 1994; 569 pp.
Dott is a professor of geology at the University of Wisconsin, and Prothero is an associate professor of geology at Occidental College. This book serves as an introductory text for courses in historical geology. After introducing the notion of historical geology, the vastness of geologic time, and various controversies in the history of geology, the authors introduce the concept of evolution and emphasize the growth of evolutionary thought since the time of Darwin. They reject creationism, which "closes the door on further inquiry" and should not be given equal time with evolution in the public schools. Both relative and absolute dating processes are reviewed, followed by speculations about the origin and early evolution of the earth and a discussion of continental drift. Much of the remainder of the book discusses the history of the earth by geologic era. A final chapter introduces environmental geology. Appendices provide a classification of living organisms and a metric conversion chart. A glossary defines important terminology. The book is exquisitely illustrated with color drawings and photographs.

Dundes, Alan (ed.). *The Flood Myth.* Berkeley: University of California Press, 1988; 452 pp.
This is a collection of 26 essays dealing with flood legends from several areas of the world. Essays at the beginning of the book deal with the biblical flood narrative and other similar legends from the Near East. Analyses of flood myths from the Americas, Australia, the Cameroon, the Philippines, Thailand, India, and Sri Lanka are also included. Two essays interpret flood myths from psychological perspectives and their symbolic content. In one, Geza Roheim suggests that flood stories are rooted in the dreams of people with full bladders who, by morning, release their urinary flood. In another, Dundes posits that there may have been an actual flood event, but that its widespread retelling is due to the symbolism of flood narratives. The final four essays deal with the biblical flood in view of the physical sciences.

Gould, Stephen Jay. *Time's Arrow, Time's Cycle.* Cambridge, MA: Harvard University Press, 1987; 222 pp.

In this, one of his finest books, Gould, who teaches biology, geology, and the history of science at Harvard, examines the nature of time. He first explores geology's discovery of "deep time," the notion that the earth is billions rather than thousands of years old. This discovery took a considerable amount of adjusting to. As Gould notes, "An abstract, intellectual understanding of deep time comes easily enough. . . . Getting it into the gut is quite another matter." He suggests that the passage of time is often perceived in two quite dissimilar ways: linearly—time's arrow, and cyclically—time's cycle. Gould believes that both metaphors of time are necessary and useful as we seek to understand earth history. He then follows these metaphors through three works of historic significance: Thomas Burnet's *Sacred Theory of the Earth,* James Hutton's *Theory of the Earth,* and Charles Lyell's *Principles of Geology.* He argues that the influence of these metaphors was more important than geological data themselves in the discovery of deep time, then uses this interpretation as a vehicle to explore the nature of scientific thought.

Chapter 7

BIOLOGY AND ANTHROPOLOGY

Life seems mysterious to many people. While they may find it easy to believe the cosmologists who say that stars evolve and the geologists who say that continents move, many people refuse to believe the biologists who say that life emerged long ago from inanimate matter and ultimately diversified into everything from slime molds to humans. The history of life and of humans is an emotionally laden topic addressed in the selected works that follow.

THEIST REFERENCES

Allen, Ronald B. *The Majesty of Man: The Dignity of Being Human.* Portland, OR: Multnomah, 1984; 221 pp.

Allen is a professor of Old Testament language and exegesis at Western Conservative Baptist Seminary. While humanism is generally held in low esteem by conservative Christians, in this book Allen makes a case for "Christian humanism"—"humanism in praise of God." Allen believes that because God created humans and bestowed his redemption upon them, humans have great dignity and importance. He believes that humans, both male and female, represent "God's finest work." The fall of humans into sin, he posits, did not erase the image of God from them. As a result of this he believes that all humans should be treated with dignity. He sees "progressive creationism" as an important advance in Christian understanding, and he considers "scientific creationism" an oxymoron. He does not seem tied to a literal six-day creation perspective. In an appendix Allen argues that the Hebrew word *bara* does not mean *creatio ex nihilo* (creation out of nothing) but "to fashion anew." His book concludes with a biblographic guide, as well as scriptural and subject indexes.

Ambrose, Edmund Jack. *The Nature and Origin of the Biological World.* New York: Halsted, 1982; 190 pp.

Ambrose is an emeritus professor of cell biology at the University of London. The first section of this book is devoted to a description of the scientific process and to a discussion of the nature of life, particularly at the molecular level. The second section addresses the complexity of life, especially in view of cellular and developmental processes. This section also provides an overview of the diversity of life, from bacteria to plants and animals. The third section tackles the topic of the history of life. Here Ambrose discusses theories about the origin of life. He also explains Darwinism and neo-Darwinism and points out some of the problems associated with these theories. He notes that genes tend to remain stable within basic kinds of organisms and also points to the paucity of transitional forms in the fossil record. He does not believe that the origin and history of life can be explained through physical and chemical principles alone. Instead, he believes that a "factor X," which he later identifies as the biblical God, was responsible for life in all its diversity and complexity.

Ashley, Benedict M. *Theologies of the Body: Humanist and Christian.* St. Louis, MO: The Pope John Center, 1985; 770 pp.

The author, a Catholic theologian, examines various views on the nature of the human body. He rejects the naturalistic notion that the body is mere "star dust," as Carl Sagan declared. He also rejects Plato's dualistic view that the soul is imprisoned in the body. For Ashley the human is "a body whose form is the soul." He suggests that God is infinitely creative because matter has infinite potentiality. Replete with philosophical and theological jargon, this book will interest readers who want to broaden their understanding of the diversity of Catholic views on matters of science and faith.

Behe, Michael J. *Darwin's Black Box: The Biochemical Challenge to Evolution.* New York: The Free Press, 1996; 307 pp.

Behe is an associate professor of biochemistry at Lehigh University. He has "no reason to doubt that the universe is the billions of years old that physicists say it is," and he finds Darwin's "idea of common descent fairly convincing." His understanding of certain biochemical mechanisms, however, leads him to believe that Darwinian evolution cannot account for "molecular machines" and biochemical pathways that he calls "irreducibly complex." Behe defines irreducibly complex systems as those in which all the components are known and in which all those components are required for function. As examples, he discusses the structure and function of cilia, the cascade of reactions responsible for blood clotting, molecular transport mechanisms, and the immune system. He also discusses biochemical pathways that, although not irreducibly complex, are so complicated that it is virtually impossible to believe they arose gradually through natural selection. He claims that nowhere in the vast literature on biochemistry is there a single detailed explanation for the gradual evolution of a complex biochemical system. Behe proposes that irreducibly complex biochemical systems arose as the result of "intelligent design" and criticizes attempts by some scientists to discount such a view out of hand. This is an entertaining, well-documented, and carefully written book. An appendix provides a simplified overview of biochemical principles.

Bergman, Jerry, and George Howe. *"Vestigial Organs" Are Fully Functional.* Kansas City, MO: Creation Research Society Books, 1990; 95 pp.

Bergman is an educator and Howe is a retired professor of biology. They begin by discussing the significance placed on vestigial organs by evolutionists and the influence these structures had on Darwin's thinking. They explain how evolutionists interpret the existence of these organs and argue that these interpretations are inadequate. They suggest that the persistent use of the vestigial organ argument by evolutionists is primarily to embarrass creationists. Even if vestigial organs existed, they argue, these structures would not counter creationist views: "The presence of vestigial organs would support only de-evolution: they would not refute design." Bergman and Howe then discuss the functions of various supposed vestigial structures, including the coccyx, human hair, tonsils, adenoids, appendix, pineal gland, nictitating membrane, Darwinian point of the pinna, male nipples, pelvic and femur bones in whales, horse splint bones, and the eyes of blind cave fish. Because "virtually all of the so-called vestigial organs are shown to have functions," Bergman and Howe argue that evolutionists cannot continue to use their existence as evidence for macroevolutionary change. The authors posit that, instead, vestigial structures are evidence for the Creator's "handiwork and design."

Birch, Charles, and John B. Cobb. *The Liberation of Life.* New York: Cambridge University Press, 1981; 353 pp.
The authors, a biologist and a theologian, set out in this book to found a new religion. This religion is based on the notion that organisms, including humans, need to be liberated from being seen as "objects to be manipulated rather than subjects that experience." While Birch and Cobb's approach is not Christian, neither is it anti-Christian. They believe in a personal God rather than an impersonal force and that God actively participates in history to the point of "even becoming human." Their God, however, is not a designer or creator, but one who urges life onward. Indeed, all life experiences an urge to integrate and experience; thus, evolution is purposive. This purposive process, in turn, should give rise to altruism, "an all-embracing love for everybody and everything."

Bowden, Malcolm. *The Rise of the Evolution Fraud: An Exposure of Its Roots.* San Diego, CA: Creation-Life Publishers, 1982; 227 pp.
In this book Bowden examines the history of the evolutionary theory from a creationist interpretive framework. For example, he believes that Darwin's physical ailments were rooted in the knowledge that what he was writing "was basically false." Also, Bowden writes that although Charles Lyell is often remembered as an opponent of biological evolutionism, he was also active in seeking to undermine the authority of the Bible. Bowden shows how Darwinism became the dominant evolutionary theory and examines the religious and sociological forces effecting this outcome. He believes that there is "an apparent balance between the evidence supporting creation and that for evolution" but that due to the intellectual domination of evolutionism, creationists today have a difficult time promulgating their views. Henry Morris wrote the foreword for the book.

Braddock, John. *Science and Nonsense.* Sussex, England: The Book Guild, 1988; 167 pp.
Braddock, a reporter for the BBC, provides a very personal critique of science and scientific culture. He believes that modern science idolizes abstract thought but denies the spirit. This, he thinks, is a reaction against medieval thought that denied the intellect but idolized the spirit. Braddock believes that acceptance of Darwinian evolution requires more faith than acceptance of creation. His own view is that evolution was nudged along by God to some unknown degree. Thus

Braddock is basically a theistic evolutionist. Despite his theism, however, Braddock denies traditional Christian doctrines such as the fall, redemption, and the second coming of Christ.

Brooks, Jim. *Origins of Life*. Belleville, MI: Lion, 1985; 160 pp.
Brooks is a Christian geochemist. In this book he presents the standard scientific theories on such topics as the origin of the universe, early earth conditions, chemical evolution, meteorites, the geologic column, plate tectonics, and the demise of the dinosaurs. He attempts to provide objective considerations of these theories and clearly differentiates what he believes to be established fact from speculations and inferences. Brooks believes in the creatorship of God but accepts the notion that the earth is very old. He discusses four views on the origin of life: supernatural creation, spontaneous generation, panspermia, and abiotic synthesis. Throughout the book he makes references to his belief that God is creator.

Clark, Gordon H. *The Biblical Doctrine of Man*. Jefferson, MD: Trinity Foundation, 1984; 101 pp.
The Trinity Foundation is a conservative Christian organization founded to correct irrationalism and point out errors in the church. This book stresses the doctrine of human nature—the meaning of the "image of God," the soul, the fall, and human depravity. Clark believes that Adam was created as a full-grown adult and disregards evolutionary models of human origins as intellectually weak. He briefly defends this view, in part, through recounting the story of the Piltdown hoax. He also believes that Adam was completely righteous and that his logical reasoning powers were perfect. Clark disdains those who disagree with his position.

Corey, Michael Anthony. *Back to Darwin: The Scientific Case for Deistic Evolution*. Lanham, MD: University Press of America, 1994; 434 pp.
This book makes a case for deistic evolution, the view that God created the universe and intentionally set it on a course toward the eventual development of humans. Corey asserts that although neo-Darwinists assume atheism, Darwin himself was a deist who believed that God created life in the beginning and that this life subsequently evolved. Corey believes that we should return to this original Darwinian perspective. He also discusses deistic evolution in relation to

human behavior. He acknowledges the fact that deistic evolutionism is not really science but rather a philosophical interpretation of the evidence from science. He believes that it is a mistake for anyone to discount the existence of supernatural powers simply because they cannot be detected.

Cosgrove, Mark P. *The Amazing Body Human: God's Design for Personhood.* Grand Rapids, MI: Baker, 1987; 202 pp.
In this book Cosgrove highlights uniquely human characteristics and the differences between humans and the anthropoid apes. For example, he posits that, except for birds, humans are the only creatures that do not need to raise their heads upward to vocalize. This capacity allows humans to talk with one another in a face-to-face position. He believes this "is important because human faces are uniquely designed for the expression of personhood." Likewise, he notes that the hymen is found only in human females. This implies that "the body human is saying that the first sexual encounter for a woman is not to be entered into lightly. Something personal and hallowed is at stake." In the same vein, Cosgrove provides several reasons for people to avoid premarital sex. He also discusses human uniquenesses in terms of childhood, mental capabilities, and death.

Cremo, Michael, and Richard Thompson. *The Hidden History of the Human Race.* Badger, CA: Govardhan Hill, 1994; 352 pp.
Cremo and Thompson are not young earth creationists, but they reject the standard evolutionary model of human origins. Instead, they contend that modern humans existed at least as far back as the Pliocene and Miocene epochs, a view rooted in the authors' appreciation of Indian Vedic literature. They support their contention by discussing evidence for out-of-place hominid fossils and the existence of stone tools in relatively early geologic strata. The authors charge that paleoanthropologists sometimes date fossils by morphological characteristics instead of by the locations of these fossils in the geologic column. They believe that some humans living today exhibit the same characteristics of ape men, and they present anecdotal evidence to support this belief. Cremo and Thompson criticize the contemporary scientific establishment which, they say, overlooks evidence unfavorable to evolutionary orthodoxy. Berkeley law professor Phillip E. Johnson wrote the book's foreword.

Croft, L. R. *How Life Began.* Durham, England: Evangelical Press, 1988; 120 pp.

Croft is a lecturer in biological sciences at the University of Salford. He begins by reviewing perspectives on the origin of life from the time of the ancient Greeks to the present. He then examines the contemporary theory of the origin of life in a primeval soup, concluding that this view is no more than a myth. He details some of the problems encountered by chemical evolutionists as they attempt to explain the origin of large biological molecules, particularly proteins and nucleic acids. He also explains that proteins are required to make nucleic acids, whereas nucleic acids are required to make proteins; thus, both types of molecules would have to have appeared simultaneously in the primitive soup. Similarly, he notes that chiral discrimination and chirality are each required for the other and argues that there is no evolutionary explanation for the origin of the triplet code of DNA. Croft is critical of Cairns-Smith's genetic takeover hypothesis which explains the "starter" form of life as occurring on clay microcrystals and points out the paucity of evidence for such a view. He notes the problems associated with the evolution of the first cell, even if one assumes the existence of biomolecules. He rejects notions of spontaneous generation and Francis Crick's hypothesis of directed panspermia as no more than science fiction. Croft's last two chapters are devoted to arguing, with William Paley, that "where there is design there must be a designer." His argument for intelligent design, just like his refutation of naturalistic evolution, is based primarily at the molecular level.

Custance, Arthur C. *Two Men Called Adam.* Brockville, Ontario: Doorway, 1983; 273 pp.

Custance is a mechanical engineer with degrees in biblical languages. His name appears on a wealth of earlier creationist literature. Here he considers a variety of questions related to what he calls the two Adams. He believes that the "First Adam" was a real human being and that Eve was literally created from Adam's rib just as the Bible reads. He discusses the notion of spirit and consciousness, the nature and purpose of the human body, men and women, death in animals and humans, the resurrection, and the meaning of heaven. He devotes considerable attention to the "Second Adam"—Jesus—and in this context examines the incarnation, virgin birth, death, and resurrection, as well as the meaning of these events to the Christian faith.

Davies, Paul. *Are We Alone? Philosophical Implications of the Discovery of Extraterrestrial Life.* New York: Basic Books, 1995; 160 pp.

In 1995 Davies became the first scientist to win the Templeton Prize for advancements in religion. In this book Davies considers the philosophical implications of extraterrestrial life. He suggests that even if humans were to discover a "single extraterrestrial microbe," our world views would be profoundly transformed. In part, he believes, such a discovery would counter the conclusion of Darwinism that life is the result of random events. Davies criticizes contemporary theories of origin of life researchers and suggests that the "complexity needed [for life] involves certain *specific* chemical forms and reactions: a random complex network of reactions is unlikely to yield life." While Davies does not rule out the possibility of creation by "a purposeful Deity," as a scientist he prefers to search elsewhere for the solution to the riddle of life origins. Rather than arising by miracle or through an improbable series of random events, Davies believes that "stupid matter has a sort of innate ability to organize itself." Indeed, he posits that "we can imagine a very long sequence of such self-organizing steps in which inert matter goes in at the top and mind comes out at the bottom." To Davies, life emerges "when matter reaches a certain level of complexity."

Davis, Percival, Dean H. Kenyon, and Charles B. Thaxton. *Of Pandas and People: The Central Question of Biological Origins.* Dallas, TX: Haughton, 1989; 166 pp.

This textbook was produced by creationists for use in public high schools. Its central theme is that the theory of intelligent design provides a more satisfactory view of the origin and history of life than does evolutionary biology. Given that the book is intended for use in public schools, religious language has been avoided. An initial chapter summarizes views presented later in the book. Later chapters examine issues surrounding the origin of life, biological adaptation, speciation, the fossil record, homologies, and biochemical diversity and similarity. In each case the authors suggest that the theory of intelligent design best accounts for the phenomena discussed. The authors do not address the perennial creationist issue of the age of the earth because design theorists are split on this issue. This interesting book pioneers a new direction in the publication of creationist textbooks.

Dinosaur Committee, Atlantic Union Conference of Seventh-day Adventists Office of Education. *A Creationist View of Dinosaurs.* Volumes 1 and 2. South Lancaster, MA: Atlantic Union Conference, 1983; 443 pp.

This two-volume curriculum guide on earth science was designed for Seventh-day Adventist elementary school teachers. "Many of us have questioned the existence of dinosaurs," notes the Dinosaur Committee in its introduction, but considering "all of the evidence presented by the thousands of specimens that have been found and placed on exhibit in museums throughout the world, there can be no doubt that these animals once lived on our earth." Cutouts, activity sheets, task cards, games, and suggested field trips are provided for teacher use. The writers of this volume rely heavily on statements by Seventh-day Adventist prophetess Ellen G. White and interpretations of dinosaur remains by Adventist pastor and dinosaur enthusiast R. F. Correia. According to the Dinosaur Committee, dinosaurs resulted from amalgamation, "the combining, or mixing, of living things to produce other living things that God did not originally create." Because dinosaurs were not part of the original creation, the Dinosaur Committee notes, "dinosaurs were among the animals that did not go into the ark."

Dodson, Edward O., and George F. Howe. *Creation or Evolution: Correspondence on the Current Controversy.* Ottawa, Ontario: University of Ottawa Press, 1990; 175 pp.

Dodson is a Roman Catholic, past professor of biology at the University of Ottawa, and author of the *Textbook of Evolution* (1952). Howe is a Baptist, a former professor of biology at Westmont College, and past president of the Creation Research Society. This book includes nearly 50 letters on the topic of evolution and creation exchanged between the two men over a five-year period. The correspondence was initiated when a letter by Dodson responding to the question "Why do the creationists win all the debates?" was printed in *Bioscience.* A wide variety of topics related to the creation/evolution controversy is discussed, from the nature of biblical authority to the Piltdown hoax. While the two men differ widely in perspective, their letters are written in a tone of mutual respect.

Duncan, Homer. *Evolution: Fact or Fantasy.* Lubbock, TX: MC International, 1986; 37 pp.

Duncan says that for 50 of his 73 years he has "been preaching the unsearchable riches of Christ." He wrote this booklet for three purposes: (1) "to show the theory of evolution is not a proven fact of science," (2) to convince believers of the Bible that their rights are violated when their children are subjected to the theory of evolution in the public schools, and (3) to demonstrate that people who insist that only evolution be taught in schools "are not as tolerant as they claim to be." He reviews the contents of a 1984 anti-creationism pamphlet distributed by the National Academy of Sciences and counters its arguments, as well as other accusations by evolutionists. He says people believe in evolution because they have been brainwashed to believe that way, because they "wish to escape the authority of the Almighty God," and because they have been deceived by Satan. Theistic evolutionists, he says, "need to repent in sackcloth and ashes." He appeals to "all Bible-believing Christians to join the battle" against evolution.

Erickson, Lonnie. *Creation vs. Evolution: A Comparison.* Poulsbo, WA: Scandia, 1993; 59 pp.
Erickson, a businessman with a background in chemical engineering, has written this little book from a creation science perspective. He states that disagreements between creationists and evolutionists arise not over scientific facts but over the interpretation of these facts. He cautions the reader to keep an open mind in the face of the evidence. He defines special creation, which he endorses, as the instantaneous origin of the basic kinds of organisms. He also defines two forms of evolution: (1) the general theory of evolution which is concerned with the origin of life and major life-forms, and (2) what he calls the "special theory" of evolution, which involves changes within species and processes of speciation. He posits that transitional forms, which are uncommon in the fossil record, provide the only evidence for the general theory of evolution. Because they are so rare, he writes, acceptance of the general theory of evolution is based on faith. He does not discuss the "special theory" of evolution.

Gish, Duane, T. *Evolution: The Challenge of the Fossil Record.* El Cajon, CA: Creation Life, 1985; 277 pp.
Gish is a biochemist and a chief spokesman for conservative creationism in America. This book is an updated version of an earlier work, *Evolution: The Fossils Say No!* (1979). Gish first argues that evolu-

tionary theory is more a matter of philosophy than of science. He then outlines and discusses predictions that emerge from what he calls the "creation model" and the "evolution model." Gish examines the question of geologic time in relation to evolution and highlights the fact that many of the early Cambrian fossils represent fully developed members of most of the modern animal phyla. He discusses the paucity of large numbers of transitional forms, which he says should be found if evolution actually happened. In this context he examines the famous horse series of fossils, *Archaeopteryx*, and so-called transitional forms of whales. He also discusses some of the fossil evidence for human evolution. In his final chapter Gish summarizes the evidence for evolution discussed earlier in the book, and he concludes that evolution is not a theory but merely a working hypothesis.

Hayward, Alan. *Creation and Evolution.* Minneapolis, MN: Bethany House, 1995; 232 pp.
Hayward is a Christian and retired government physicist from England. He explains that his purpose is to provide ordinary people with a guide to the creation/evolution controversy. Part I contains a series of objections to Darwinism, not from creationists, but from non-Darwinian evolutionists. Hayward notes that Darwinism is far from universally accepted by evolutionists outside of England and the United States. Part II summarizes evidence bearing on the age of the earth. Hayward takes a dim view of young earth creationism and its associated model of flood geology. Part III examines what the Bible has to say about creation. Hayward argues against theistic evolution and in favor of a literal, historic fall of Adam. In the last chapter he posits that, while the earth is very old, convincing evidence for large-scale evolutionary change is nowhere to be found.

Hefner, Philip. *The Human Factor: Evolution, Culture and Religion.* Minneapolis, MN: Fortress, 1993; 317 pp.
Hefner, a systematic theologian at the Lutheran School of Theology, Chicago, sets out to make "conversation between theology and the sciences." He begins by stating his theory of humankind as the "created co-creator," then discusses the nature and purpose of theory in general, suggesting that a theory does not have to be correct to be useful. He then goes on to discuss the human experience in the context of nature, freedom, and culture. Lastly, he attempts to bridge his ideas with those of Christian tradition and theology. Hefner argues

from a naturalistic point of view, suggesting that humans are the products of the interaction between genes and culture. Likewise, revelation comes only by a study of nature by humans, and values, morals, and culture have arisen through evolutionary processes. Religious belief and ritual are important but are continually reinvented by humans to meet the needs of time and place. According to Hefner, the ultimate purpose of humankind will be worked out by evolution.

Johnson, Phillip E. *Darwin on Trial.* Washington, DC: Regnery Gateway & InterVarsity Press, 1991; 196 pp.
Johnson is a contentious Berkeley law school professor who specializes in the analysis of logical arguments. His book has received much publicity and has been widely reviewed. Johnson, who is a Christian but not a young earth creationist, provides a scathing critique of the Darwinian revolution, particularly the philosophy of naturalism which often accompanies Darwinian viewpoints. He does not doubt the reality of the microevolutionary process but is critical of the assumption that microevolutionary mechanisms can account for macroevolutionary change. He criticizes Darwinists for ruling out the possibility of design and purpose in nature and deplores what he considers to be a misuse of information and power to maintain the evolutionary dogma in society.

Kautz, Darrel. *The Origin of Living Things.* Milwaukee, WI: Self-published, 1988; 68 pp.
Kautz describes himself as a "lifelong . . . student of the Bible." This book examines four ways of conceptualizing the origin of life. His primary question is: "How did the universe, earth, vegetation, fish, birds, animals, and human beings come into existence?" He provides four possible answers: (1) simply say we don't know, (2) by chemical evolution, (3) by theistic evolution, or (4) by biblical creation. On the basis of his discussion of scientific data and natural laws, Kautz dismisses the first three options and selects the fourth as the most reasonable answer.

Lester, Lane P., and Raymond G. Bohlin. *The Natural Limits to Biological Change.* Grand Rapids, MI: Zondervan, 1984; 207 pp.
Lester and Bohlin are Christian biologists who are comfortable with the notion of biological change, as long as that change is understood to be limited in extent. They first summarize, then critique, funda-

mental concepts of genetics, ecology, neo-Darwinism, and punctuated equilibrium. They argue that postulated mechanisms of genetic change, such as gene duplication, neutral mutations, and mutations of regulatory genes, are inadequate to account for wholesale evolution. They believe that the original "prototypes" (created kinds) contained sufficient DNA variability to diversify into a plethora of species under the influence of recombination and natural selection. Organisms with similar developmental pathways, they suggest, share common ancestries back to these prototypes. Lester and Bohlin hope that biologists will seek to identify these prototypes through comparative studies of developmental pathways.

Lubenow, Marvin L. *Bones of Contention: Creationist Assessment of Human Fossils.* Grand Rapids, MI: Baker, 1992; 295 pp.
Lubenow is a conservative creationist who believes that the fossil record falsifies the theory that humans evolved from lower primates. After introducing the topic of paleoanthropology, Lubenow examines the fossil evidence for Neanderthal man, *Homo erectus*, and archaic *Homo sapiens.* Most of his effort is spent on *Homo erectus*, which he considers to be crucial for a satisfactory understanding of human paleoanthropology. On the basis of assigned radiometric ages, he argues that fossils of modern *Homo sapiens* were preserved in rocks of the same age as those containing fossils of *Homo erectus*, *Homo habilis*, and *Australopithecus*. Consequently, he discounts the notion of a human evolutionary tree. He explains why ideas about human origins are important to Christians, discusses the relationship of the Big Bang theory to human history, argues that Genesis is a very old book for which Moses served as a redactor or editor, and critiques non-literal readings of Genesis. The book concludes with an appendix on radiometric dating.

McCann, Lester J. *Blowing the Whistle on Darwinism.* Waconia, MN: Self-published, 1986; 124 pp.
McCann is a professor of biology at the College of St. Thomas. He begins by reviewing the history of evolutionary theory and its detractors, through the time of Darwin and the neo-Darwinists to the present. McCann views Darwin as an amateur scientist and questions the accolades he continues to receive today. He sees Darwinism as a hindrance to progress in biological science. McCann alleges that the spontaneous origin of life through abiogenesis would have been a

violation of the second law of thermodynamics. He is critical of the concept of natural selection because it has no way to control energy and create order. He also criticizes the notion that mutations can cause evolution because they are mistakes and mistakes cannot improve coded information. Darwinists, he writes, are generally mechanists who view life at the chemical level; life, however, cannot be explained simply on the basis of physics and chemistry. McCann posits that three things are necessary for the production of any structure: materials, energy, and intelligence. He concludes by suggesting that scientists and non-scientists who are concerned about the impact of Darwinism on society should join efforts in their common cause.

Pannenberg, Wolfhart. *Anthropology in Theological Perspective.* Philadelphia, PA: Westminster, 1985; 552 pp.
Pannenberg is a professor of systematic theology at the University of Munich. In this scholarly tome the eminent theologian attempts to integrate theology and the science of anthropology. He notes that early "proofs" for the existence of God were based in "natural theology," but this endeavor eventually gave way to a more anthropological outlook. This new perspective examined God from the perspective of human existence and experience. Christianity can effectively confront its atheistic critics only from an anthropological perspective, he believes, because the critics also argue from an anthropological perspective. Pannenberg argues that if the biblical God is the creator of both human and natural reality, then theology should be able to extend the dimensions of scientific anthropology. Thus, he calls for an appropriation of anthropology by theology. This theological anthropology, he believes, will be characterized by two fundamental themes: that humans embody the image of God and that humans experience sin. Pannenberg also deals with many other theological themes in this comprehensive study.

Parker, Gary. *Creation: Facts of Life.* Colorado Springs, CO: Master Books, 1994; 215 pp.
Parker, a biologist, is a former evolutionist who converted to creationism following a three-year intellectual and spiritual struggle. He first provides a series of evidences for creation from biochemistry, embryology, and ecology. He then takes up the topic of design and contrasts the views of Darwinian evolution with those of creationism. He notes that Charles Darwin was a careful scientist, but he dis-

putes the commonly held view that Darwin was the originator of the theory of natural selection. That distinction, writes Parker, rightly belongs to creationist Edward Blyth, who viewed the process of natural selection as happening to a fallen creation. Parker accepts the reality of natural selection but believes this process allows for only limited change within the original created kinds. God, he opines, created the original variation upon which much of natural selection works. Natural selection that works on variation produced by mutations, he argues, results mainly in disease-causing organisms and birth defects. Parker is a young earth creationist who interprets the fossil sequence in terms of ancient life zones and the rising waters of the flood.

Peacocke, Arthur. *God and the New Biology.* San Francisco, CA: Harper & Row, 1986; 197 pp.
Peacocke, a biochemist and Anglican priest, shares his reflections on the implications of molecular biology for our understanding of God and nature. Peacocke first establishes the significance of reductionist thought to progress in science. He then discusses the variety of ways individuals have sought to relate science and faith. His own particular position allows for extensive cosmic and biological evolution, including the spontaneous origin of life. Human nature, he believes, is the product of natural processes and matter is capable of organizing itself into progressively more complex forms.

Peth, Howard A. *Blind Faith: Evolution Exposed.* Frederick, MD: Amazing Facts, 1990; 188 pp.
Peth begins by asking if evolution is scientific. He responds that "Evolutionary theory is no less religious and no more scientific than special creation." While he says that no one can deny organisms have changed since creation, this change has been within limits—microevolution occurs but not macroevolution. He discusses the inability of mutations and natural selection to produce new kinds of organisms and denies that comparative anatomy and vestigial organs support evolutionary theory. According to Peth, life could never have arisen by chemical evolution; moreover, the sudden appearance of organisms in the fossil record and the existence of missing links refutes the claims of evolution, as do the so-called hominid fossils. Peth says that people believe in evolution because of the authority of science, not because of the weight of the evidence. He is critical of theistic evolution and the day-age theory of creation, both of which

compromise what he believes to be the truth of a literal six-day creation.

Pitman, Michael. *Adam and Evolution.* Grand Rapids, MI: Baker, 1984; 268 pp.
The author, a biology teacher with training in science and the classics, rejects the premise that the world and its inhabitants could have resulted from evolution. By contrast, "this book advocates a grand and full-blooded creation." Despite this assertion, Pitman is critical of fundamentalist creationism. He believes the biblical account of creation to be mythical and seems to accept the view that the world and life are very old. He appears to favor progressive creationism.

Raven, Charles. *Science, Religion and the Future.* Harrisburg, PA: Morehouse, 1994; 125 pp.
This book, one of the volumes in Morehouse's Library of Anglican Spirituality, is a reprinted version of a series of Raven's lectures first published in 1943. Susan Howatch, editor of the series, first introduces readers to Raven, who became Regius Professor of Divinity at Cambridge University in 1932. Raven, educated in both biology and theology, devotes the first part of the book to the development of science and its interactions with Christianity. He considers the role of inductivism in science but shows that Bacon was not the only source of the so-called scientific method. He also examines the impact of evolutionary theory on nineteenth century British society and the conflicts it engendered. Raven, a strong advocate of evolutionism in the tradition of Teilhard de Chardin, would like to see a "new Reformation" which understands the origin of human consciousness, the life of Jesus, and development of community in evolutionary terms.

ReMine, Walter James. *The Biotic Message: Evolution Versus Message Theory.* St. Paul, MN: St. Paul Science, 1993; 538 pp.
ReMine works in the area of "pattern recognition, signal processing, biomedical engineering, and radio communications." He is also an amateur magician. He believes that the theory of evolution is merely an illusion. By contrast, his "message theory" predicts that "Life was designed to look like the work of a single designer." According to ReMine, message theory predicts and explains "life's key pattern," a pattern best studied by "discontinuity systematics" which does not assume common ancestry among organisms. The combination of dis-

continuity systematics and message theory he calls "Creation Systematics . . . the scientific study and explanation of biological creation." He says that the paucity of ancestral forms in the fossil record is evidence of the truth of his view, as is the existence of common biomolecules like DNA and RNA. He believes that the existence of many structures thought by evolutionists to be the result of convergence, similarities between apes and humans, and the occurrence of imperfections, oddities, and vestigial structures in organisms give evidence of a single designer who is attempting to communicate his existence through the signature left in these structures. Although ReMine is clearly a creationist, for the most part he avoids using religious language, and his references are primarily from the standard evolutionary literature.

Rusch, Wilbert H. *The Argument—Creationism vs. Evolutionism.* Norcross, GA: Creation Research Society Books, 1984; 86 pp.
In this book Rusch, a founder of the Creation Research Society and a frequent contributor to creationist literature, examines the opposing claims of creationism and evolutionism. He opines that Charles Darwin's popularity was based, in part, on his provision of a model that seemed to liberate people from responsibility to a personal God. Rusch discusses the fact that early in life Carolus Linnaeus believed in the fixity of species but that later in life he allowed for variation within limits—a position commensurate with the views of modern creationists. Rusch counters the arguments of prominent evolutionists and provides quotations to show that even evolutionists have serious doubts about the ability of their theory to account for the fossil record and the complexity of life. He reviews the evidence for fossil humans and concludes that modern humans were contemporaneous with supposed ancestral forms, including *Australopithicus.* Appendices include an annotated bibliography, pertinent quotations, a list of creationist organizations, and a reprint of the "Humanist Manifesto."

_____. *Origins: What Is at Stake?* Kansas City, MO: Creation Research Society Books, 1991; 73 pp.
Rusch's daughter, Joanne Rusch, a systems architect, has written an introduction to this book in which she opines that creationism should be taught in both the home and the school. Her father begins the body of the text by defining his terms. A believer in the absolute inerrancy of the Bible, the elder Rusch is particularly opposed to the

theory of macroevolution. Any accommodation to this view, he believes, destroys Christian views of God, scripture, miracles, the nature of man, original sin, and the historicity of the flood. He discusses three evidences which, for him, place macroevolutionary theory and uniformitarianism in question: the occurrence of cyclothems, prostrate fossil tree trunks, and insects preserved in amber. He agrees with Philip Gosse that God would have created many objects with an appearance of age. Appendices examine the meaning of taxonomy and the nature of the geologic column, list biblical references to creation, suggest other readings, and provide a set of quotations.

Saint, Phil. *Fossils That Speak Out: Creation vs. Evolution.* Phillipsburg, NJ: P & R Publishing, 1985; 119 pp.
Saint is a missionary and artist who obtained a bachelor's degree in anthropology from Wheaton College. Much of this book is based on what Saint learned from Alexander Grigolia, who chaired the department of anthropology at Wheaton when Saint was a student there. Saint contrasts creationism, which teaches "a sudden fall from an originally perfect state," with humanism, which teaches "a long, gradual climb out of a bestial past." He believes that theistic evolution discredits the inspiration of scripture. After briefly recounting Darwin's views, Saint discusses reasons to doubt evolution, including the purported tracks of pre-flood giants and evidence for design. He also discusses the political and social misuses of evolutionary ideas. He reviews the hominid fossil evidence and concludes that convincing evidence for human evolution is nowhere to be found. This book is heavily illustrated by Saint's well-executed cartoons.

Silvius, John E. *Biology: Principles and Perspectives.* Dubuque, IA: Kendall/Hunt, 1994; 443 pp.
Silvius, an evangelical Christian, is a professor of biology at Cedarville College. This book was written as a text for students taking introductory biology. Unlike most biology texts, this one is written from a biblical point of view—the first chapter is "A Scriptural Perspective of Life." Throughout the book references to the Bible are commonplace. In addition to many of the standard introductory biology topics, Silvius discusses creationism, scientific creationists, and evolution.

Sippert, Albert. *Evolution Is Not Scientific: 32 Reasons Why.* Mankato, MN: Sippert, 1995; 448 pp.

Lutheran pastor Sippert provides this updated and expanded version of his earlier book, *From Eternity to Eternity.* A critic of macroevolution, Sippert begins by discussing human history from a biblical perspective. He then addresses a variety of topics related to evolutionary theory including the Big Bang, evolutionary hoaxes, missing links, dinosaurs, thermodynamics, and radiometric dating. According to Sippert, evolutionary theory is rooted in atheistic humanism and cannot be supported by available scientific evidence. Beginning with the 1925 Scopes trial, Sippert reviews famous legal battles over the teaching of evolutionism and creationism in America's public schools. The book concludes with a postlude, "A Tribute to Our Creator-Redeemer God."

Sunderland, Luther D. *Darwin's Enigma.* El Cajon, CA: Master Book Publishers, 1984; 177 pp.

Sunderland is an aerospace engineer and a frequent public speaker on the topic of creationism. This book provides an examination of the fossil record and its implications for the theory of evolution. Sunderland first notes the fact that Charles Darwin was puzzled by the lack of evidence from the fossil record favoring "gradual and continuous evolution." Sunderland discusses the fossil animals preserved in the Cambrian strata and observes that they seem to appear suddenly and without antecedents. Moreover, these fossils seem to indicate the existence of significant gaps between the major groups of animals. Sunderland then discusses fossil vertebrates, arguing that evidence for transitional forms is lacking. Data from genetics and embryology, he says, create more problems for evolutionists than solutions. The theory of punctuated equilibria, he argues, is an inadequate attempt to explain away the gaps in the fossil record. He concludes that Darwinism lacks a scientific foundation—that what the evidence really shows is that life appeared abruptly and in many forms and that organisms were subsequently buried by worldwide, catastrophic events.

Thaxton, Charles B., Walter L. Bradley, and Roger L. Olsen. *The Mystery of Life's Origin: Reassessing Current Theories.* New York: Philosophical Library, 1984; 228 pp.

All three of the authors of this book have Ph.D.'s—Thaxton in chemistry, Bradley in materials science, and Olsen in geochemistry.

Their book consists of a critique of the theory, first proposed by Oparin and Haldane in the 1920's, that life originated through a process of chemical evolution. After introducing the topic of chemical evolution, the authors examine the assumptions regarding the nature of the primordial soup and the earth's primordial atmosphere, the experiments designed to synthesize prebiotic monomers, the thermodynamics of living systems in relation to life origins, and the nature of postulated protocells in relation to modern cells. Among other topics, they discuss the low yields of chemicals in origin of life experiments, the instability of intermediate compounds, and the difficulty of developing functioning biopolymers from the monomers produced in these experiments. They conclude that it is unlikely that life originated by chemical evolution. In the final chapter they present five alternative means by which life could have arisen, including divine creation. The book is written in an objective style and is well referenced.

Verbrugge, Magnus. *Alive: An Enquiry into the Origin and Meaning of Life*. Vallecito, CA: Ross House Books, 1984; 159 pp.
Verbrugge, a retired urologist, here examines the nature and history of the theory that life originally arose spontaneously through abiogenesis; he argues that this theory is based more in philosophical materialism than in scientific data. Although he believes that life is more than a complicated set of physical and chemical interactions, he rejects the animistic notions of Hobbes, Leibnitz, Newton, Buffon, and others that living matter contains spiritual attributes. Recent supporters of abiogenesis such as Oparin and Monod, Verbrugge claims, are animists as well. Verbrugge describes the complexities of the cell and its molecules, demonstrating that cellular processes are anything but random. In a discussion of DNA and its function, Verbrugge notes that enzymes are necessary for the production of DNA but that DNA orchestrates the production of these enzymes. He argues that scientists who favor abiogenesis force their data into their preconceived models about how life evolved. He favors the perspective of Herman Dooyeweerd that each level of complexity exhibits irreducible characteristics, precluding the gradual evolution of living cells. Verbrugge claims that materialistic scientists, who oppose traditional religious explanations for the origin of life, substitute their own religious philosophy, a naturalistic one. Thus, Verbrugge believes that anyone who addresses the question of the origin of life does so from a religious perspective, either traditional or naturalistic.

von Fange, Erich A. *Helping Children Understand Genesis and the Dinosaur.* Syracuse, IN: Living Word Services, 1992; 208 pp.

This book is written to help parents, pastors, teachers, and children understand the relationship between dinosaurs and the Genesis account of creation. Von Fange, an elementary school teacher, once lived in the Alberta badlands where dinosaur remains are abundant. He has examined dinosaur skeletons both in the field and in museums. He provides evidence for Christians that dinosaurs really did exist, explains what life was like when they were alive, and provides the meanings of dinosaur names and suggestions on how to talk with people who do not believe that dinosaurs existed. He also discusses dinosaur tracks, the formation of fossils, and carbon dating. He notes that dinosaur fossils raise many questions left unanswered by either the Bible or science.

White, A. J. Monty. *Wonderfully Made.* Durham, England: Evangelical Press, 1989; 128 pp.

White, a creationist chemist, believes that evolution serves as an excuse for nontheists to reject God and his law. He believes that the Bible should be seen as an owner's manual for our moral character and physical bodies, which he believes are "fearfully and wonderfully made." He also holds that scripture provides an authoritative source of information on the history of life. After explaining why one's perspectives on origins is important, White highlights some of the major differences between creationist and evolutionist worldviews. He discusses several problems for evolutionary theory, including gaps in the fossil record, problems in paleoanthropology, the falsity of embryonic recapitulation, and the lack of truly vestigial organs. He is critical of theistic evolution and believes that the biblical story of Adam and Eve is historical. He posits that God reveals himself through both scripture and the creation.

Wilder-Smith, A. E. *The Scientific Alternative to Neo-Darwinian Evolutionary Theory.* Costa Mesa, CA: TWFT Publishers, 1987; 198 pp.

In this book Wilder-Smith applies concepts of information theory to questions about the origin of life. He argues that no combination of matter, time, and energy could produce life—information is a necessary fourth factor. The author provides an extensive discussion of the second law of thermodynamics and posits that self-replicating living systems could not arise out of random processes without the input of extrinsic information. Wilder-Smith categorizes information as of two types: actual and potential. Single bits of organized matter can, indeed, arise spontaneously, he says, but these bits contain only potential, not actual, information. Information to organize single bits, he believes, could have been from some dimension outside our space-time universe. Wilder-Smith is critical of both chemical evolutionary theory and the view that life came from outer space on meteorites. He laments the lack of academic freedom displayed by administrative bodies' handling of the cases of three non-Darwinian scientists: astronomer Sir Fred Hoyle, physicist Robert Gentry, and biologist Dean Kenyon. While he favors a non-materialist outlook on the origin of life, he is critical of the "internecine fighting" among creationists.

Wright, Richard T. *Biology Through the Eyes of Faith.* San Francisco: Harper and Row, 1989; 298 pp.
Wright teaches biology at Gordon College. This volume is one of eight supplemental textbooks produced by the Christian College Coalition. This book is intended as a supplement to introductory biology textbooks. Wright develops his discussion around four revolutions in biological science: Darwinian, biomedical, genetic, and environmental. He points out that everyone approaches science with a basic philosophy about how science operates. He believes that Genesis 1 is God's word in nontechnical language about how things appear—it is not concerned with how God created. Wright discusses four approaches to relating science and scripture: concordism, substitutionism, compartmentalism, and complementarism. He opts for the last of these approaches. He also deals with the problem of life origins, the Darwinian revolution, biomedical ethics, the ethics of biotechnology, and the importance of Christian environmental stewardship. In the last chapter Wright challenges readers to an active faith, one committed to justice, peace, and care of our fellow humans and the environment. This is a clearly written, well-organized book.

NONTHEIST REFERENCES

Angela, Piero, and Alberto Angela. *The Extraordinary Story of Human Origins.* Translated from Italian by Gabriele Tonne. Buffalo, NY:

Prometheus, 1993; 328 pp.
Piero Angela hosts television programs on science and is a best-selling Italian author. Alberto Angela has participated in paleontological research in Zaire and Tanzania. This book explores the evolutionary history of humans on the basis of the available fossil evidence. The authors follow the story chronologically, from the earliest fossil hominids to modern humans. Discussions of the fossil evidence are interspersed with hypothetical reconstructions of everyday life of the various evolutionary stages. The book contains four appendices, one of which examines the question of how fossils are dated. The authors point out the tentativeness of any conclusion based on the very limited fossil evidence available.

Barbieri, Marcello. *The Semantic Theory of Evolution.* New York: Harwood Academic, 1985; 200 pp.
Barbieri reviews the history of ideas on evolution, examines the evidence suggestive of a primitive reducing atmosphere for the earth, discusses the biochemical evidence for evolution, and critiques several theories of life origins. He finds none of the previous theories compelling, so presents his own—the "Ribotype Theory." In this view "ribosoids," molecules that contain ribose sugar such as RNA and ATP, were most responsible for the evolution of early life. He rejects the dualistic "genotype>phenotype" view of cell function, favoring instead a "trinity of genotype> ribotype>phenotype." He hastens to add that his is not a simple restatement of the "Central Dogma" of molecular biology, which is commonly stated as "DNA>RNA> proteins," the RNA being messenger RNA. By contrast, Barbieri sees ribosomal RNA as the crucial molecule. In his view, ribosoids became nucleosoids, nucleosoids became heterosoids, and heterosoids became microkaryotes. Microkaryotes, in turn, diverged to become prokaryotes and microeukaryotes, the cell types from which all modern life forms evolved.

Barlow, Connie (ed.). *Evolution Extended: Biological Debates on the Meaning of Life.* Cambridge, MA: MIT Press, 1994; 333 pp.
Barlow, who was trained in zoology, has drawn on the writings of many authors, including Francisco Ayala, John T. Bonner, Charles Darwin, Richard Dawkins, Stephen J. Gould, John Greene, Thomas Henry Huxley, Peter Medawar, Jacques Monod, Henry Morris, David Raup, George Gaylord Simpson, Brian Swimme, and Edward O. Wil-

son, to produce this interesting and unusual volume. In addition to the short quotations and full-length essays by these and other well-known authors, Barlow includes illustrations, poetry, and her own reflections, making this a very personal work. Although she has been influenced by New Age perspectives of the earth, particularly James Lovelock's Gaia hypothesis, she includes many contrasting viewpoints, thereby allowing the reader to come to his or her own conclusions. The book is well referenced and indexed.

Berra, Tim M. *Evolution and the Myth of Creationism: A Basic Guide to the Facts in the Evolution Debate.* Stanford, CA: Stanford University Press, 1990; 198 pp.
Berra, a zoology professor at Ohio State University, is concerned with the perceived threat that scientific creationism poses for the educational process. He discusses the nature of science, tenets of creationism, the concept of evolution, the fossil record, radiometric dating, the explanatory power of evolutionary theory, cosmic evolution, abiogenesis, the emergence of the major groups of organisms, and the evolution of humans. He critiques some of the major creationist claims against evolution and discusses the creationist movement within the context of major social trends. This is a well-written, nicely illustrated book designed to demonstrate the force and validity of evolutionary theory for the general reader.

Brace, C. Loring. *The Stages of Human Evolution.* Fifth edition. Englewood Cliffs, NJ: Prentice Hall, 1995; 371 pp.
This successful textbook by Brace, a prominent paleoanthropologist at the University of Michigan, has been published since 1967. Brace intends for this book to focus "on the major trends that have characterized the course of human evolution and the circumstances that contributed to the changes that took place." He begins with a chapter on methods of interpreting human evolution, from religious as well as scientific perspectives. He posits that "there need be no conflict between science and religion," noting that "Many scientists are deeply religious people" committed to a variety of traditional denominational belief systems. He suggests, however, that his book is not for Christians who read the Bible literally. Brace examines pre-Darwinian views of human history, important paleoanthropological discoveries, and the evolutionary principles through which these discoveries are typically interpreted. He then discusses how geologists date the past

and provides a brief overview of vertebrate evolution through to the origin of primates. This is followed by chapters on culture and ecology, the emergence of australopithecine diversity, the development of members of the genus *Homo*, and the appearance of Neanderthal and modern humans. An epilogue briefly considers the future of humankind and warns readers that interpretations about human history may change as new discoveries are made.

Brooks, D. R., and E. O. Wiley. *Evolution as Entropy: Toward a Unified Theory of Biology.* Chicago, IL: University of Chicago Press, 1986; 335 pp.
The authors of this volume posit an approach to macroevolutionary theory in terms of thermodynamic principles. They note that standard evolutionary theory has failed to (1) "come to grips with the underlying causal laws of chemistry and physics," (2) successfully integrate developmental biology into its theoretical framework, (3) "provide a rationale for the existence of higher taxa," and (4) "provide . . . a robust explanation of the relationship between form and function in evolution." They believe that it is impossible to resolve these shortcomings "within the current theoretical framework." They see species "as living systems partly closed in terms of information and cohesion but open in terms of energy" and the generation of new species as occurring with increases in entropy. This is accomplished, they write, by increasing organization through accession of "new microstates while retaining . . . historical access to initial microstates." Brooks and Wiley favor sudden rather than gradual change, but they see this change as "an inherently nonviolent, accommodating process."

Cairns-Smith, A. G. *Seven Clues to the Origin of Life: A Scientific Detective Story.* New York: Cambridge University Press, 1985; 131 pp.
Cairns-Smith begins by completely discounting the standard theory of chemical evolution as an explanation for the origin of life. He believes there is no possibility that life could have originated in this fashion, regardless of the amount of time available. He then introduces his own, now well-known, theory that organic chemicals maintained the stability of clay crystals and gave them the ability to replicate. Eventually, the organic chemicals completely replaced the clay crystals, thus creating the first cells. The book is written with the

general reader in mind, and the author artfully frames the problem and solution in the form of a Sherlock Holmes-type mystery.

Crick, Francis. 1981. *Life Itself.* New York: Simon and Schuster, 1981; 192 pp.

Nobel laureate and biophysicist Francis Crick is best known for his codiscovery (with James Watson) of the structure of DNA. In this book, Crick addresses the question of how life originated. He first explains why he believes that life could not have evolved spontaneously on Earth. He notes, for example, that the probability of a protein forming spontaneously in the correct sequence is roughly one chance in 10^{260}, a staggering implausibility considering the fact that there are only around 10^{80} particles of matter in the total universe. He also points out that living systems must be able to replicate, facilitate energy conversions, and transfer information from generation to generation. He then explains how difficult it would be for these and other features to develop spontaneously. Consequently, he argues that life arose, probably from RNA, on some other planet where conditions were more favorable to the origins process. Moreover, life advanced on that planet to a very high form which then intentionally spread living particles, perhaps as bacteria, throughout the universe on specially constructed rockets. Crick discusses how these rockets may have been designed and how they could have successfully entered the earth's atmosphere. This is a highly speculative book by one of the twentieth century's best-known scientists regarding one of life's greatest mysteries.

Dawkins, Richard. *The Blind Watchmaker.* New York: Norton, 1986; 318 pp.

This book provides a response to the "argument from design" articulated most famously by William Paley in 1802. According to Paley, a watch implies the existence of a watchmaker; by analogy, creatures imply the existence of a creator. Dawkins, an Oxford don, acknowledges the strength of this argument but believes that it has now been falsified by Darwinism. According to Dawkins, "the only watchmaker in nature is the blind forces of physics, albeit deployed in a very special way." Dawkins argues his point using game models of natural selection and expositions of genetic information storage and replication. A committed gradualist, Dawkins claims that punctuationists like Stephen Jay Gould are really gradualists at heart. Sharp-witted writer that he is, Dawkins has a penchant for examining old ideas in new ways.

_____. *River Out of Eden: A Darwinian View of Life.* Science Masters Series. New York: Basic Books, 1995; 172 pp.
In this book, Dawkins sets out to "accord due recognition to the inspirational quality of our modern understanding of Darwinian life." He seeks to accomplish this by explaining how the diversity of life we see can be explained from an evolutionary perspective. The "river out of Eden" is a metaphor for DNA's river of information, now exhibiting 30 million branches (species). He explains how speciation occurs through geographic isolation and the subsequent development of reproductive isolating mechanisms; traces human life back to a Mitochondrial Eve who lived a couple hundred thousand years ago in Africa; argues that complexity and beauty are not the products of design but of gradual evolution; asserts that natural selection and the universe as a whole have no purpose; and reviews the history and future of life, from its spontaneous origin to its possible, though unlikely, colonization of other parts of the universe.

Deamer, David W., and Gail R. Fleischaker (eds.). *Origins of Life, the Central Concepts.* Boston, MA: Jones and Bartlett, 1994; 431 pp.
This anthology contains 46 papers published from 1908 to 1992 on the origin of life. The articles are arranged in five sections: "The Early-Earth Environment," "Prebiotic Chemistry," "Self-assembly of Supramolecular Systems," "Energetics of Life's Origins," and "Bioinformational Molecules." As the editors note in the foreword, this book provides a "smorgasbord of ideas" on early influences on the life origins—everything from cometary impacts to the "RNA world." The authors are well aware of the speculative nature of their topic. The articles are of a technical nature and are reprinted as photoreproductions of the original publications.

De Duve, Christian. *Vital Dust: Life as a Cosmic Imperative.* New York: Basic Books, 1995; 362 pp.
Christian de Duve, emeritus professor of medicine at Rockefeller University, won the Nobel Prize in 1974 for his work on the structural and functional characteristics of cells. In this book de Duve attempts to reconstruct the history of life, from the chemical evolution of the first cells through to the emergence of the human mind. He be-

lieves that life arose through purely naturalistic means, although we still do not understand many of the steps involved. He presents available evidence, as well as many theories and speculations about the evolution of life. In the last several chapters de Duve takes up more philosophical issues such as human free will, ethics, purpose, and meaning. He writes clearly but readers without a reasonably good grasp of biochemistry may find some of his arguments difficult to follow. An annotated bibliography and glossary of terms make this book a useful reference tool.

Dennett, Daniel C. *Darwin's Dangerous Idea: Evolution and the Meanings of Life.* New York: Simon & Schuster, 1995; 586 pp.
Dennett is the director of the Center for Cognitive Studies at Tufts University. This book addresses the meaning of reality when Darwinian thinking is followed to its logical conclusions. He has written the book "for those who agree that the only meaning of life worth caring about is one that can withstand our best efforts to examine it. Others are advised to close the book . . . and tiptoe away." Part I examines philosophical paradigms that were popular before Darwin, introduces natural selection as an algorithmic process, and shows how revolutionary Darwinism really is for our worldviews. Part II defends Darwinism against such notions as punctuated equilibrium, panspermia, genetic transmission of acquired traits, and directed mutation. Part III addresses the implications of Darwinism for our understanding of language, culture, morality, and human diversity. Dennett concludes "that Darwin's idea is a universal solvent, capable of cutting right to the heart of everything in sight" but that it leaves us "with stronger, sounder versions of our most important ideas."

Denton, Michael. *Evolution: A Theory in Crisis.* London: Burnett, 1985; 368 pp.
Denton is a molecular biologist at the Prince of Wales Hospital, New South Wales, Australia. Although he views Charles Darwin and his work with respect and recognizes the significance of natural selection for microevolutionary and speciation processes, he argues that natural selection cannot explain macroevolutionary change. He begins by reviewing the shift from a biblical view of earth history to Darwinian and post-Darwinian views. He then discusses how natural selection is involved in the formation of new species. He notes that the typological views of early taxonomists are at odds with the concept of evolu-

tionary change, which assumes a continuum of life-forms over time. He evaluates the concept of evolutionary homology but explains that homologous structures sometimes develop via different embryonic routes. He sees the paucity of transitional forms in the fossil record as a very significant problem for macroevolution. Five chapters near the end of the book are devoted to biochemical problems for macroevolution and the origin of life. Denton dismisses the Genesis creation story as an ancient myth but concludes that "the Darwinian theory of evolution . . . is the great cosmogenic myth of the twentieth century." He offers no substitute for either view.

Edey, M. A., and D. C. Johanson. *Blueprints: Solving the Mystery of Evolution.* Boston, MA: Little, Brown, 1989; 418 pp.
This book traces the history of evolutionary thought from the time of Linnaeus to the present. Three chapters review Charles Darwin and Alfred Russel Wallace's discovery of evolution by natural selection. A large section of the book is devoted to the history of hereditary concepts, particularly of DNA as the primary genetic material and as a central component of the evolutionary process. This is followed by a chapter on paleoanthropology and human evolution. A final chapter posits that the human intellect may have evolved to a point at which the civilization it has created is generating more problems than can be readily solved. Edey and Johanson take a dim view of creationism, suggesting that creationists misinterpret and ignore evidence from the natural world. They posit that if God is assumed to be responsible for nature, no one will bother to examine nature scientifically.

Eldredge, Niles. *Reinventing Darwin: The Great Debate at the High Table of Evolutionary Theory.* New York: Wiley, 1995; 244 pp.
Eldredge curates the invertebrate collections at New York City's American Museum of Natural History. He is best known for the theory of punctuated equilibrium, which he developed in collaboration with Stephen Jay Gould during the early 1970's. Eldredge and Gould devised their theory to more accurately reflect the fossil record, which often evidences abrupt species transitions followed by long periods of stasis. By contrast, more traditionalist-minded population geneticists postulate that species changes occur gradually over long periods of time. This book details the nature of this fundamental disagreement between paleontologists and geneticists, with Eldredge explaining why he favors the punctuated equilibrium model. Although this book

deals with some of the intricacies of speciation models and evolutionary theory, it is written for the nonspecialist.

Fox, Sidney. *The Emergence of Life: Darwinian Evolution from the Inside.* New York: Basic Books, 1988; 208 pp.
Fox, a biochemist known for his experimentation in the realm of life origins, explains why he believes proteins evolved before nucleic acids and how resultant microspheres developed into living cells. He has few followers in the origin of life community but blames this on the "channeled thinking" of neo-Darwinists and his fellow chemists. Fox's theory depends a great deal on the self-organization potential of molecules. In contrast to many scientists and philosophers, he believes that nature is somewhat deterministic.

Godfrey, Laurie R. (ed.). *What Darwin Began: Modern Darwinian and Non-Darwinian Perspectives on Evolution.* Boston, MA: Allyn and Bacon, 1985; 312 pp.
In her preface Godfrey notes that while evolutionary theory is a hotly debated topic today, "the very existence of plausible and actively debated alternatives signals not the death of Darwinism but the life and vitality of modern evolutionary theory." This book consists of a collection of essays that focus on the controversy in evolutionary science. The first section, which provides a backdrop against which later sections can be assessed, consists of four essays with historical and philosophical perspectives on evolutionary theory. The second section contains five essays challenging aspects of neo-Darwinism, with discussions of topics such as neutral selection theory, punctuated equilibrium theory, the role of chance in evolution, and the theory of the inheritance of acquired characteristics. The third section contains two essays on evolution and the public, with one focusing on scientific creationism and the other on the presentation of evolution in public museum exhibits. The last section examines four controversial issues related to origins: the origin of the cosmos, the origin of life, the origin of multicellular animals, and the origin of humans. A glossary defines technical terms used by the various authors.

Gould, Stephen Jay. *Hen's Teeth and Horse's Toes.* New York: Norton, 1983; 413 pp.
Gould is a professor of geology, biology, and the history of science at Harvard University. He is one of the twentieth century's most sig-

nificant voices in science, particularly in evolutionary biology and geology. This volume is Gould's third collection of essays (after *Ever Since Darwin* [1977] and *The Panda's Thumb* [1980]); most were previously published in his monthly "This View of Life" column in *Natural History*. Articles in this collection show how nature's oddities support evolutionary theory, examine the views of past scientists, explore processes of adaptation and development within the context of life's history, speculate as to who was behind the great Piltdown hoax, discuss racism, and look at the nature of extinction. In addition, one essay discounts the claim by creationists that evolution is only a theory, another recounts a visit by Gould to Dayton, Tennessee, site of the famous Scopes Trial of 1925, and a third laments the impact of creationism on America's schoolchildren.

_____. *The Flamingo's Smile*. New York: Norton, 1985; 476 pp.
This is Gould's fourth published collection of essays. In the prologue he writes: "*The Flamingo's Smile* is about history and what it means to say that life is the product of a contingent past, not the inevitable and predictable result of simple, timeless laws of nature." Many of Gould's favorite themes are explored: evolutionary explanations for "quirky" structure and function in nature, the meaning of statistical trends in view of historical contingency, the myth of progress, human equality and the dangers of eugenics, historical Darwinism, and mass extinctions. This collection also includes an examination of Philip Henry Gosse's book *Omphalos* (1857), which championed the view that God created a very old-looking world only a few thousand years ago, and a look at the problems and logic of flood geology in the early nineteenth century and today.

_____. *An Urchin in the Storm: Essays About Books and Ideas*. New York: Norton, 1987; 255 pp.
This book consists of a collection of reviews written by Gould for the *New York Review of Books*. He makes little reference to creationism in these reviews, except to briefly express animus, particularly toward flood geology. But four of his essays deal directly with evolutionary themes, while another two address time and geology. He also hammers away against biological determinism, the topic of his earlier book *The Mismeasure of Man* (1981). He examines the lives of four biologists and offers thoughts on the importance of reason in today's world.

_____. *Wonderful Life.* New York: Norton, 1989; 347 pp.
Gould turns orthodox evolutionary theory on its head in this fascinating book which describes and interprets the Burgess Shale fossil fauna of Yoho National Park, British Columbia. Discovered in 1909 by Charles Walcott, this site contains Cambrian fossils representing most of today's animal phyla, plus a variety of other unusual animals that Gould refers to as "weird wonders." According to Gould, the traditionalist Walcott "shoehorned" all the Burgess animals into modern animal phyla, even though some of them clearly did not fit. These misfits represented several ancient animal phyla, he writes, phyla that appeared early on during an initial and rapid explosion of body types, then disappeared. Despite the tendency to think of evolution as resulting in an increasing cone of diversity through time, Gould says that the fossil record at every level supports the hypothesis that initial diversities of body plans are followed by decimations—in other words, *decreasing* cones of diversity. He posits that contingency has played an important role in the history of life—that if the history of life could be replayed, it would follow a different pathway. This book is thought-provoking, well-referenced, and beautifully illustrated with pen-and-ink drawings of the Burgess fauna.

_____. *Bully for Brontosaurus.* New York: Norton, 1991; 540 pp.
This is Gould's fifth, and he believes "best," collection of essays to date. Again, his main themes are evolutionary change, the oddities of nature, the personalities of science, and the nature of history. He explores dinosaurs, adaptation, art and science, Australian animals, probability, and implications from the *Voyager* planetary probe. Gould devotes five essays to the evolution/creation controversy: a revisionist version of the famous interaction between Thomas Huxley and Biship Wilberforce in 1860, reflections on the relationship between Genesis and geology, an exploration of the roots and nature of William Jennings Bryan's creationism, a postmortem on the infamous misidentification of a pig molar as a tooth from an ancient American ape-man and the misappropriation of this story by creationists, and an analysis of the U.S. Supreme Court's 1987 decision to strike down the last state creationist statute.

_____. *Eight Little Piggies.* New York: Norton, 1993; 479 pp.
Billed as his "most contemplative and personal," this sixth collection of Gould's essays addresses his standard potpourri of themes. In

several pieces he focuses on environmental deterioration and extinction. He considers the evidence for evolutionary transitional forms by examining vertebrate digits, the tail bends in ichthyosaurs, the conversion of reptile jaw bones to mammalian middle ear ossicles, and the relationship between the swim bladders of fish and the lungs of terrestrial animals. He introduces several historical personages and considers the vagaries and evolution of human nature, revisions and extensions to Darwinism, and large-scale patterns of evolution. He endorses William Paley's argument from design, but concludes that the "designer" is natural selection, not a divine creator. While he does not endorse Archbishop James Ussher's six-thousand-year chronology for the world, Gould nevertheless defends the quality of the seventeenth century prelate's scholarship and takes issue with the notion that science progresses by beating down religion.

_____. *Full House: The Spread of Excellence from Plato to Darwin.* New York: Harmony, 1996; 244 pp.
Gould offers this book as a companion volume to his much acclaimed *Wonderful Life* (1989). In *Wonderful Life* he used the Burgess Shale fossil fauna as a vehicle to discuss the importance of contingency in the history of life. *Full House* crusades for another of Gould's favorite themes: that the history of life is not defined by any grand trend such as progress. In other words, evolution did not inevitably lead to the human species which occupies "no preferred status as a pinnacle or culmination. Life has always been dominated by its bacterial mode." Variation in systems, he believes, is more important than complexity. What appear to be trends are simply changes in variation. Gould illustrates this theme by discussing the disappearance of 0.400 hitting in baseball. Baseball hitters have not become progressively worse, but baseball as an entire system has changed, making it more difficult for batters to hit. What appears to be a declining trend is simply the result of a reduction in the variation of batter success as the entire game has changed. By the same token, changes in organisms through time are often interpreted as trends, but Gould argues that such changes are simply shifts in the variability of highly variable systems, a principle understood long ago by Charles Darwin. While Gould does not deny "the increased complexity in life's history," he rejects the notion that this phenomenon is a significant feature of most lineages. Thus, any illusion of progress is "a purely incidental consequence" of changing variation.

_____ (ed.). *The Book of Life: An Illustrated History of the Evolution of Life on Earth.* New York: Norton, 1993; 256 pp.
This book is a cooperative effort which involves the knowledge and talent of both scientists and artists. Richly illustrated chapters introduce readers to contemporary views regarding the evolution of life on earth. Chapter 1 examines the history of attempts to understand the history of life. Chapter 2 introduces concepts of geologic time and the fossil record. The third chapter speculates on how life arose four billion years ago. Subsequent chapters address the development of fish, the transition from legless fish to four-footed amphibians, the origin and history of dinosaurs, the emergence of mammalian diversity, and the emergence of modern humans from ancient primates. The writing, illustrations, and graphics of this book are superb.

Greenstein, George. *The Symbiotic Universe: Life and Mind in the Cosmos.* New York: William Morrow, 1988; 271 pp.
Greenstein, an Amherst College professor of astronomy, is struck by the incredible mystery of the existence of life in the universe. In the first 11 chapters he discusses the special properties of the cosmos that make it habitable to life. In the next three chapters he explains why he thinks the universe is habitable. From quantum mechanical theory he shows that subatomic particles exist only if observed by a conscious mind; given that all structures of the universe consist of subatomic particles, none of them could exist without conscious observation. But, of course, conscious minds derive from brains, which are also made of particles. Thus, minds exist only because of particles and particles exist only because of minds—mutual interdependence. Greenstein opines that belief in the supernatural hinders scientific progress. Ultimately, he believes, everything in the universe will be explained on the basis of physical law.

Hitching, Francis. *The Neck of the Giraffe: Where Darwin Went Wrong.* New York: Ticknor and Fields, 1982; 288 pp.
Hitching is not a creationist, but in this book he roundly criticizes Darwinian evolutionary theory. He begins by pointing out the large number of gaps in the fossil record, which he believes precludes a process of gradual evolution. He then examines genetic mechanisms of change and concludes that there seem to be natural limits to the extent of change. The large-scale changes required by Darwin's general theory, he says, cannot be accounted for on the basis of currently

understood genetic mechanisms. Hitching raises the question of the origin of life and suggests that a scientific understanding of this unique event is probably beyond the realm of possibility. Certainly, he argues, the chance of the spontaneous formation of life is vanishingly small. Hitching discusses the origin of complex structures like hair, feathers, and the eye and concludes that none of these could have arisen gradually by natural selection. He examines alternative theories of origins, including creation, which he ultimately rejects. Hitching also discusses evolutionary time and criticizes the reductionist tendencies of modern biologists. He believes that a better theory of evolution than Darwin's will eventually be promoted. He recounts the stories of Piltdown Man and Peking Man and suggests that Darwinists have misused evidence to support their theory. The last chapter is devoted to the life of Charles Darwin.

Hoffman, Antoni. *Arguments on Evolution: A Paleontologist's Perspective.* New York: Oxford University Press, 1989; 274 pp.
Hoffman is a European paleontologist who believes that, despite its critics, neo-Darwinism continues to offer a lot to evolutionary theory. In the first chapters of the book, Hoffman dismisses creationism, examines neo-Darwinism, discusses the importance of the fossil record, and points out some of the problems associated with the reconstruction of history. He then turns his attention to the problem of macroevolution and attempts to provide a neo-Darwinian perspective on this process. In this view, Hoffman is critical of the concepts of both punctuated equilibrium and species selection. Next, he discusses "megaevolution," large-scale patterns of change through time interrupted by extinction events. He believes that megaevolutionary patterns are simply the accumulated results of microevolutionary events. Hoffman's views provide an interesting contrast with those of many of his contemporary paleontologists who are skeptical that microevolutionary processes can account for evolutionary patterns.

Hollar, David W. *The Origin and Evolution of Life on Earth.* Pasadena, CA: Salem, 1992; 235 pp.
Hollar is a professor at Rockingham Community College, Wentworth, North Carolina. This book is an annotated bibliography that references major evolutionary works from the late 1800's to the early 1990's. Entries are organized into chapters that focus on general studies, evidence supporting evolution, evolutionary controversies, cos-

mology, primordial life, prokaryotes, viruses, eukaryotes, debates over evolutionary descent, the fate of the dinosaurs, the development of intelligence in mammals, human origins and evolution, and the search for extraterrestrial intelligence. Hollar's book serves as an excellent bibliographic tool for anyone searching for information on evolutionary biology.

Hoyle, Fred. *Evolution from Space.* Wales: University College Cardiff Press, 1982; 30 pp.
In this lecture delivered at the Royal Institution, astronomer Sir Fred Hoyle, one of the twentieth century's foremost cosmologists, explains why he thinks life could not have originated on earth. Like biophysicist Francis Crick, Hoyle believes that the conditions on earth are simply unfavorable for the spontaneous origin of life. Living things must have originated elsewhere and come to earth as passengers on meteorites. In light of this proposal, Hoyle speculates that certain microscopic structures in meteorites are actually fossil bacteria. Moreover, he suggests that microbes continue to shower the earth from space today. Indeed, some of the viral epidemics we experience on earth from time to time may be caused by extraterrestrial viral invaders. Like Crick's earlier book, *Life Itself* (1981), Hoyle's lecture provides a highly speculative account of the appearance of life on earth.

_____, and N. Chandra Wickramasinghe. *Evolution from Space.* London: J. M. Dent and Sons, 1981; 176 pp.
Hoyle and Wickramasinghe believe that Darwin acquired his views on natural selection from Edward Blyth, whom Darwin failed to acknowledge, and that Darwinism retains its popularity more for sociological factors than for compelling scientific reasons. They ask, for example, why bacteria and fruit flies would develop the exact mutations needed to deal with specific types of radiation damage. In view of this skepticism, they reject the theory that life evolved from nonliving matter on earth. Instead, they posit that genetic material enters our atmosphere continuously from outer space. This material may be in the form of DNA fragments, whole bacteria, and even insect eggs. Benevolent life-forms, they believe, are intentionally spreading these bits of life throughout the universe. For evidence, they appeal to the fact that bacteria are found in low concentrations as high as ten miles above the earth's surface. They assert that these concentrations, while

low, are still much higher than one would expect on the basis of patterns of air movement from low to high altitude. Hoyle and Wickramasinghe believe that "God" consists of the entire universe.

Janabi, T. H. *Clinging to a Myth: The Story Behind Evolution.* Indianapolis, IN: American Trust Publications, 1990; 166 pp.
Janabi, who works in the fields of robotics and artificial intelligence, takes the reader through his personal, often idiosyncratic viewpoints on evolutionary theory. Janabi is critical of evolution on metaphysical and on mathematical grounds. He draws on the behavioral sciences, Muslem philosophy, and the theory of general relativity to support his views. He argues against abiogenesis and believes that evolution cannot account for characteristics that are uniquely human.

Kauffman, Stuart. *At Home in the Universe.* New York: Oxford University Press, 1995; 321 pp.
Kauffman is a biologist at the Santa Fe Institute who specializes in the self-organization of systems. This book explores the notion that life is the result of both chance and necessity. Although Kauffman does not doubt that Darwinian evolution plays a significant role in living systems, he rejects the view that chance governs all. He believes that no one has previously developed a viable theory for the origin of life. Then, through the use of computer simulations, he seeks to show that complex, self-replicating, autocatalytic systems can develop in a chemical soup under the right conditions. He uses chaos theory to show that cells and ecosystems exist along functional boundaries. He provides explanations for the Cambrian "explosion" of life-forms and the great Permian extinction. At the end of the book he uses his theory of self-organization to explain economic, political, and social systems.

Kitcher, Philip. *Abusing Science: The Case Against Creationism.* Cambridge, MA: MIT Press, 1982; 213 pp.
Kitcher wrote this book to serve as "a manual for intellectual self-defense" against proponents of creationism, particularly representatives of the Moral Majority and the Institute for Creation Research. He first summarizes contemporary evolutionary theory. He then critiques a variety of creationist assertions. For example, to the creationist assertion that evolution is not science, he replies that evolutionary theory produces testable hypotheses and predictions, a hall-

mark of science; to the argument that evolution is implausible because mutations are harmful, not helpful, Kitcher counters that whether a mutation is harmful or helpful depends on the genetic construction and environment of the organism; to the criticism that the fossil record fails to show gradual transitions from one group to another, he responds that some such transitions *are* found, although they are uncommon due to the incompleteness of the fossil record. Kitcher, then, turns creationist arguments back on creationism itself, arguing, for example, that neither is creationism science nor is it supported by evidence from the natural world. He closes with the assertions that misquotation "has become a way of life" for creationists; that Christianity has fostered many evils including anti-Semitism, the Inquisition, and witch-burning; and that one use for creationism in the schools might be as a good example of pseudoscience.

Leakey, Richard, and Roger Lewin. *Origins Reconsidered: In Search of What Makes Us Human.* New York: Doubleday, 1992; 375 pp.
Leakey, the prominent paleoanthropologist, and Lewin, the award-winning science writer, have teamed up to produce a nicely written, beautifully illustrated book that explores theories regarding the origin and evolution of humankind. Discoveries by Leakey and other paleoanthropologists are highlighted, followed by an extended discussion of the implications of these discoveries for aesthetics, psychology, and human genetics. The authors believe that humans are not genetically predisposed to aggressive behavior, a tendency that developed after the appearance of cities, agriculture, and accumulated wealth. They believe that humans are not special creations but merely end products of chance evolutionary processes. The human mind, they believe, is a remarkable and powerful human attribute, but its existence is due to purely evolutionary processes.

Løvtrup, Søren. *Darwinism: The Refutation of a Myth.* London: Croom Helm, 1987; 469 pp.
Swedish embryologist Løvtrup rejects Darwin's theory of evolution by natural selection as a myth. He does allow that the natural selection of micromutations accounts for microevolutionary processes within populations that result in insignificant changes, but larger-scale change, he believes, must be due to macromutations in developmental control genes. His book provides a review of various evolutionary theories and the history of the personalities that promoted

them. Løvtrup is particularly critical of Darwin and notes that most of Darwin's friends and supporters did not entirely accept his views. Løvtrup is more favorably impressed with Lamarck's contributions to evolutionary theory and thinks that it was really Lamarck, not Darwin, who did more to popularize the concept of evolution. Løvtrup blames Herbert Spencer for the development of social Darwinism. The last chapter expresses the author's belief that progress in biology has been crippled by undue adherence to a Darwinian paradigm. Løvtrup's unorthodox views are controversial but thought provoking.

McGowan, C. *In the Beginning: A Scientist Shows Why Creationists Are Wrong*. Buffalo, NY: Prometheus, 1984; 208 pp.
This book is a critique of creationist views published in two sources: Henry Morris' *Scientific Creationism* (1974) and Duane Gish's *Evolution: The Fossils Say No* (1973). McGowan posits that evolutionary theory is supported by the fact that the oldest fossils are of the simplest organisms and that different types of organisms appear at different levels of the geologic column. He favors the theory of punctuated equilibrium as a means to explain the paucity of gradual evolutionary sequences of fossils. He discusses the evolutionary implications of *Archaeopteryx* and the evolution of horses, as well as the postulated transitions from fish to amphibians, amphibians to reptiles, and reptiles to mammals. He also discusses the fossil evidence for human evolution. He cites work by Urey, Miller, and Fox to support his view that life arose in a primordial soup. Viruses, he says, represent a form of matter between life and non-life. He is highly critical of creationist views on numerous topics including the laws of thermodynamics, living fossils, Noah's flood, radiometric dating, meteoritic dust, and cosmology.

McKinney, Michael L. *Evolution of Life: Processes, Patterns, and Prospects*. Englewood Cliffs, NJ: Prentice Hall, 1993; 415 pp.
McKinney teaches evolutionary biology at the University of Tennessee. His goal for this undergraduate textbook "is to introduce the principles of evolution to the non-scientist." McKinney begins by recounting the history of evolutionary thought and its impact on religion, philosophy, ethics, business, and politics. This is followed by an exposition on evolutionary theories of the universe, life, and humans, and a review of the history of life in the context of the products and patterns of physical, biological, and human evolution. Mc-

Kinney concludes with chapters on the impact of extinctions and biotechnology on future evolution, as well as some personal and social implications of evolutionary theory. He writes that many natural scientists continue to believe in a "Supreme Being" because science has not explained where the natural laws responsible for life came from. While "a personified God may exist," he says, "there is no physical evidence of it thus far. Given this, the most immediate outlook is self-reliance." According to McKinney, this means that humans must look out for their own best ecological well-being, as well as develop self-understandings and personal philosophies that are rooted in a knowledge of evolutionary theory.

Mayr, Ernst. *Toward a New Philosophy of Biology: Observations of an Evolutionist.* Cambridge, MA: Harvard University Press, 1988; 564 pp.

Ernst Mayr is one of the chief architects of the twentieth century's "evolutionary synthesis," his best-known contribution being the theory of allopatric speciation. This large tome is a collection of 28 essays by the famed evolutionist, covering such topics as philosophy, natural selection, adaptation, Darwin, diversity, species, speciation, macroevolution, and history. He is strongly opposed to teleological and theistic perspectives, believing that humans must reject Judeo-Christian values in order to respond favorably to the pressures of natural selection. He also argues against the possibility of finding intelligent extraterrestrial life and discusses the origins of human ethics. Mayr's insights into evolutionary theory and its history make for valuable reading.

Morgan, Elaine. *The Scars of Evolution: What Our Bodies Tell Us About Human Origins.* New York: Oxford University Press, 1994; 196 pp.

Morgan supports the theory that humans evolved from apes living in an aquatic environment. Human features such as bipedalism, fat distribution, lack of fur, a descended larynx, menstrual cycling, tears, and face-to-face copulation are discussed as "scars" remaining as a result of the existence of early hominids in an aquatic environment. This view contrasts with the more popular notion that modern humans evolved from early hominids living on the African savannah. This well-written book is designed for the nonspecialist.

_____. *The Descent of the Child: Human Evolution from a New Perspective.* New York: Oxford University Press, 1995; 197 pp.
The Descent of the Child is a sequel to Morgan's earlier book, *The Scars of Evolution: What Our Bodies Tell Us About Human Origins* (1994). Both works interpret aspects of human structure and function as telltale signs of the aquatic environments in which our primate ancestors lived. This book provides a detailed description of human development from conception to puberty. In addition to evidences favoring the aquatic theory mentioned in Morgan's previous book, her sequel highlights several additional human features that she interprets as having originated in water. For example, she postulates that the large brain size of humans evolved in response to selective pressures exerted by three-dimensional aquatic locomotion, that human speech arose as a consequence of voluntary breath control and the descended larynxes of swimming primates, and that rates of human growth and development reflect adaptations to the buoyancy of water. The iconoclastic thesis of this book should provoke vigorous debate on the part of readers interested in the biological history of humankind.

Morrill, Thomas A. *Evolution as Growth of One Earth-Organism.* Tallahassee, FL: Self-published, 1995; 200 pp.
Morrill, a retired high school biology teacher, considers himself a "naturalist, poet, scientist, . . . idiot and genius." This book contains his reflections on evolution and religion. Although Morrill assumes the reality of evolution, he rejects Darwinian natural selection as too random. In his view, evolution is a self-directed process tending toward greater complexity through cooperation, rather than as result of competition. He believes that intelligence was present in the primeval molecular interactions that led to the first cells. In his view, evolution reached perfection in the creation of humans. All other organisms evolved for our benefit. Morrill does not see evolution as a branching bush but as a ladder with humans forming the top rung. This idiosyncratic book ranges far outside the realm of contemporary scientific and religious thought.

Price, Peter W. *Biological Evolution.* Fort Worth, TX: Saunders, 1996; 429 pp.
Price teaches biology at Northern Arizona University. This book was written as an undergraduate textbook in evolutionary biology. The first section begins with an examination of the nature of science, fol-

lowed by an introduction to the history of Darwinism and a discussion of neo-Darwinian principles. The second section, "Macroevolution," introduces species concepts, mechanisms of speciation, the origin and subsequent evolution of organisms, adaptive radiation, extinction, the fossil record, human evolution, and biological classification. The third section, "Microevolution," examines the genetics of evolutionary processes, non-Darwinian evolution, evolutionary rates, and coevolution. A conclusion addresses the question, "Is the theory of evolution adequate?" Price answers that there are no scientific alternatives available. Well-executed black-and-white drawings illustrate each chapter. A glossary appears after the last chapter.

Rachels, James. *Created from Animals: The Moral Implications of Darwinism.* New York: Oxford University Press, 1991; 245 pp.
In this well-written, carefully argued book, Rachels challenges both theists and many practitioners of mainline science to question their assumptions about the nature of human morality. Darwin, he writes, dissolved the spiritual and moral as well as the biological distinctions between humans and animals. Thus, if humans are not special in a biological sense, neither are they special in a moral sense. Rachels rejects not only the design argument and theistic evolutionism but also sociobiological theory which seeks to provide an evolutionary explanation for the existence of human ethics. He considers the few true practitioners of Christian love as aberrations from the norm. He believes that ethics can be developed only because life has value—not to anyone or anything except the individual, human or animal, that experiences life. On this basis he rejects the notion that humans should have dominion of other species, although he does not believe that all species share the same moral status. This book attempts to push Darwinian theory to its ultimate moral implications.

Radnitzky, Gerard, and W. W. Bartley III. *Evolutionary Epistemology, Rationality and the Sociology of Knowledge.* LaSalle, IL: Open Court, 1987; 475 pp.
This anthology emerged out of a series of discussions, correspondences, and conferences on the topic of evolutionary epistemology. The essayists consist of a historian, a psychologist, and several distinguished philosophers, including Sir Karl Popper. The first third of this volume examines the logic underlying explanations of the evolutionary process. The last two-thirds of the book evaluates the theory

of rationality and discusses the sociology of knowledge. Specific topics addressed include the adequacy of Darwinism to account for evolutionary change, the alleged tautology of argument to explain natural selection, the evolutionary development of instinct, and the nature of the relationship between mutation and selection in the adaptive process. The arguments are technical and will be most accessible to readers trained in philosophy.

Ridley, Mark. *Evolution.* Boston, MA: Blackwell, 1993; 670 pp.
Ridley is a professor of biology at Emory University. This book is intended as an introductory text in evolutionary biology. Ridley believes that the "theory of evolution is outstandingly the most important theory in biology." Part One examines the history of evolutionary theory, provides a genetic basis for understanding evolutionary process, reviews some of the evidence for evolutionary change, and introduces the concepts of variation and natural selection. Part Two explores evolutionary genetics, including the genetics of natural selection, molecular evolution, quantitative genetics, and genome evolution. Part Three examines the notion of natural selection in more detail and discusses the principle of adaptation. Part Four looks at species and speciation, classification, the construction of phylogenies, and evolutionary biogeography. Part Five discusses rates of evolution, macroevolutionary processes and trends, and extinction. Ridley rejects scientific creationism but believes that "No important religious beliefs are contradicted by the theory of evolution, and religion and evolution should be able to coexist peacefully in anyone's set of beliefs about life."

Rifkin, Jeremy. *Algeny.* New York: Viking, 1983; 298 pp.
Rifkin, an environmentalist gadfly and outspoken opponent of genetic engineering, believes that we are moving from an age of environmental manipulation led by the industrial revolution to an age of genetic manipulation. According to Rifkin, "People create cosmologies to sanction their behavior" and Darwinian evolution was a spatial cosmology created by the industrial revolution. The new age of biotechnology will create a new cosmology. Rifkin reviews some criticisms of Darwin's theory and suggests that a new theory of evolution is emerging, one based in temporal, not spatial, reality. The new theory sees organisms differing from one another on the basis of differing temporal programming, by which Rifkin means differing

minds. "All species . . . are bundles of knowledge, and each species is distinguished by its intelligence; that is, the speed with which it is able to utilize knowledge to control its own future."

Rolston, Holmes, III (ed.). *Biology, Ethics, and the Origins of Life.* Boston, MA: Jones and Bartlett, 1995; 248 pp.
Rolston is a professor of philosophy at Colorado State University. This anthology resulted from a conference held at CSU in 1991. Rolston introduces this volume by noting that the origin of life and the origin of humans are "critical points of intense biological and philosophical interest." The writers of this volume "move from the beginning to the present, trying to understand ethics in its interaction with biology, evaluating origins to discover the nature in, and the nature of, our duties." First, molecular biochemist Thomas Cech claims that the mechanistic origin of life does not deny its specialness as an entity to be valued; writer Dorion Sagan and biologist Lynn Margulis see the ethical responsibilities of humans as expanding in view of the Gaia hypothesis; paleontologist Niles Eldredge feels that humans, as evolutionary latecomers following several mass extinctions, should promote ethics as a matter of self-interest and survival; philosopher Michael Ruse sees human cooperation and ethics as survival tools developed through natural selection; geneticist Francisco Ayala holds that ethics arose as an evolutionary byproduct of human intelligence and that behaving ethically may or may not have survival value; philosopher Elliott Sober believes that the human brain, a product of natural selection, is the creator of ethics, which is itself freed of the constraints of selection; theologian Langdon Gilkey assumes that humans are more than products of biological history, and that they evidence a higher morality unexplained by science; and biologist Charles Birch advocates a postmodernist theology that sees God as a "persuasive influence . . . to which individual entities respond."

Ruse, Michael. *Darwinism Defended: A Guide to the Evolution Controversy.* Reading, MA: Addison-Wesley, 1982; 356 pp.
Ruse is a prominent philosopher of biology and critic of creation science. He also served as a prosecution witness in the Arkansas "Scopes II" trial in 1981. Parts I-III of his book address the history and nature of Darwinism. Ruse believes that the claims of science, unlike those of religion, can be empirically evaluated. He discusses the biological evidence favoring the theory of evolution and

unabashedly sings "the praises of Darwin," defending neo-Darwinism against what he considers to be the passing fads of punctuated equilibria and cladistics. He believes that neo-Darwinism has an even "more glorious future" than its "proud past." Part IV addresses the topics of human evolution, sociobiology, and ethics. Part V consists of a searing condemnation of creationism, a movement Ruse considers "totally, utterly, and absolutely wrong . . . ludicrously implausible . . . a grotesque parody of human thought . . . a downright misuse of human intelligence . . . [and] an insult to God." By contrast, he says "I love and cherish Darwinian evolutionary theory."

_____. *Taking Darwin Seriously.* New York: Basil Blackwell, 1986; 296 pp.
The first part of this book provides an overview of Darwinian evolution, followed by a critical evaluation of earlier attempts to derive knowledge and morality from Darwinism. Ruse then provides his own rationale for morals within the context of neo-Darwinism. He rejects the notion of a God-based morality, because he says a moral God would not have allowed evil. Human morality, according to Ruse, resulted from group co-adaptation that promoted individual genetic fitness. Our perception of an objective human morality is merely "a collective illusion foisted upon us by our genes."

_____ (ed.). *But Is It Science?* Buffalo, NY: Prometheus Books, 1988; 406 pp.
Michael Ruse has assembled here a diverse collection of papers on the issue of creationism, particularly its philosophical implications. The book begins with an essay by Ruse about his experience as an expert witness at the Arkansas "Scopes II" trial in 1981. This is followed by four parts. Part One addresses nineteenth century issues and includes documents from the period plus more recent essays written about the time. Part Two explores contemporary evolutionary theories including views of natural selection, gradualism, punctuated equilibrium, saltation, and speciation. Part Three looks at the history and perspectives of modern creationism, including a paper by creationist debater Duane Gish. Finally, Part Four contrasts the opinions of Ruse with those of two other philosophers, Larry Laudan and Philip Quinn. Specifically, criteria for what is appropriately called "science" are examined, especially in relation to Judge Overton's decision in the Arkansas trial.

Shapiro, Robert. *Origins: A Skeptic's Guide to the Creation of Life on Earth.* New York: Summit Books, 1986; 322 pp.
Shapiro, a chemist at New York University, critiques many of the contemporary theories for the origin of life, along with the biblical story, as myths—authoritative accounts of the facts which are not to be questioned. He suggests that many origin of life theorists in science are so blinded by their particular viewpoints, they have lost their capacity to evaluate scientific evidence objectively. While Shapiro is just as critical of scientific creationist approaches to life origins, he posits that "if all other explanations should fail, in the end we will have no option but to accept the idea of supernatural forces." Until that point is reached he believes "we must look for rational ways of accounting for the data." While this book deals with technical subject matter, it is written for a popular audience.

Sober, Elliott (ed.). *Conceptual Issues in Evolutionary Biology: An Anthology.* Cambridge, MA: MIT Press, 1986; 725 pp.
This anthology contains 35 previously published essays that address issues at the border between philosophy and evolutionary biology. The 27 contributors include some of the chief architects of modern evolutionary science and philosophy—people like Richard Dawkins, Stephen Jay Gould, Richard C. Lewontin, Ernst Mayr, Michael Ruse, John Maynard Smith, and Edward O. Wilson. Their perspectives differ dramatically. Essays composing the first section of the book address foundational concepts in evolutionary genetics and population biology; the second section looks at the misunderstood concept of evolutionary fitness; the third set of essays attempts to identify the basic unit of natural selection (is it the gene, the individual, or the group?); the fourth and fifth sections examine the concepts of adaptation, optimization, and function, and attempt to develop an evolutionary "teleology without vitalism"; the sixth section focuses on the problem of reductionism in genetics, which during the latter half of the twentieth century shifted from transmission genetics to molecular biology; and the final section consists of a lively debate among pheneticists, cladists, and evolutionary taxonomists as they attempt to define a meaningful species concept. This volume provides a wealth of information on modern evolutionary thought.

Stebbins, G. Ledyard. *Darwin to DNA, Molecules to Humanity.* San Francisco, CA: Freeman, 1982; 491 pp.

Stebbins is a biologist and professor emeritus at the University of California, Davis. He wrote this book to serve as a nontechnical introduction on evolution. Part One introduces the concept of evolution, Darwinism, neo-Darwinism, genetic variation, microevolution and macroevolution, and the fossil record. Part Two considers the origins of life, organelles, cells, sex, and cellular diversity, and reviews the theories of how plants and animals, including nonhuman primates, evolved. Part Three focuses on human biological and cultural evolution, with considerations of the hominid fossil record, the evolution of mental ability and culture, and theories of human sociobiology. Stebbins also examines history from an evolutionary perspective and speculates about the future of the human species.

Taylor, Gordon Rattray. *The Great Evolution Mystery.* New York: Harper & Row, 1983; 277 pp.
Taylor is a science writer and chief science advisor to BBC Television. He is firmly committed to the concept of naturalistic evolution but highly critical of neo-Darwinism as an adequate mechanism for evolutionary change. Here he reviews evidence that seems to refute the theory of evolution by natural selection. For example, he points to the coordinated series of reactions that make blood in humans and the intricacies of the photosynthetic process, then asks how natural selection could possibly generate such complicated, interrelated systems. He also examines the fossil record, noting that paleontologists find scant evidence for the major evolutionary pathways they have postulated. Taylor fails to propose an adequate evolutionary theory of his own, but he seems to favor the concept that organisms have innate tendencies toward greater complexity. Although it has not yet done so, eventually, he believes, science will discover an adequate mechanism for evolutionary change.

Tyler, Donald E. *Originations of Life from Volcanoes and Petroleum: A Scientific Theory Opposed to Evolution.* Portland, OR: Ryan Gwinner, 1983; 128 pp.
Tyler is a retired physician who states that it is "time that the theory of evolution was discarded." He seeks to replace evolutionism with his own "theory of multiple originations and creations." In this view, hot gases inside the earth reacted to form petroleum, which oozed to the earth's surface. Here the petroleum reacted with water and other chemicals to produce DNA, RNA, and proteins which, under just the

right conditions, formed cells. Some of these cells were germ cells. Plants grew from some of these cells and, along with anaerobic bacteria, began to produce oxygen. Once oxygen levels increased, some of the germ cells grew into invertebrate animals. Land plants eventually developed, then land animals, all from petroleum-generated germ cells which were continuously produced in volcanic regions. According to Tyler, species are fixed from the time of their initial appearance. Also, the various races of humans emerged independently of one another in areas with petroleum seeps. For example, he suggests that Native Americans "probably originated in Southern California where there are numerous petroleum seepages."

Ward, Peter Douglas. *On Methuselah's Trail: Living Fossils and the Great Extinctions.* San Francisco, CA: W. H. Freeman, 1992; 212 pp.
This book is concerned with two prominent concepts derived from a study of the fossil record: first, mass extinctions, global events during which large percentages of the earth's biota have gone extinct; and second, living fossils, organisms that made it through mass extinctions and continue to survive today. Ward, a paleontologist, also discusses Darwin's theory of evolution in relation to the sudden appearance of diverse animal forms in the Cambrian rocks. Ward suggests that the theory of punctuated equilibrium provides the best explanation for this evidence. He cautions against reading too much about ancestral relationships from the skeletal features of fossil organisms without knowing what their soft parts were like. He intersperses his discussion with interesting accounts of paleontological field trips, which provide the reader with a sense of discovery.

Weiner, Jonathan. *The Beak of the Finch: A Story of Evolution in Our Time.* New York: Knopf, 1995; 332 pp.
Weiner, a science writer and a former editor of *The Sciences*, here recounts two decades of research by Peter and Rosemary Grant on the famous finches of the Galapagos Islands, birds noted by Charles Darwin during his visit there in 1835. He begins by describing Daphne Major, the tiny island on which the Grants have done most of their work and the finches that populate this and other islands of the archipelago. Thirteen species of these birds have been described, differing primarily in bill shape and size. The Grants have brought us a long way toward understanding the evolutionary mechanisms by which

these birds diversified. Specifically, they have demonstrated a web of interactions involving the birds and their environment. These interactions create selective pressures that result in measurable changes in the birds' bills. Weiner applies the results of the Grants' work to other biological examples, and he makes a compelling case for the action of natural selection in the living world. The last chapter briefly considers the claims of creationists in light of the finch research. Exquisite pen-and-ink drawings add interest and charm to this book, which won Weiner a Pulitzer Prize.

Wills, Christopher. *The Wisdom of the Genes: New Pathways in Evolution.* New York: Basic Books, 1989; 351 pp.
Wills is a professor of biology and member of the Center for Molecular Genetics and the University of California, San Diego. In this book he examines how recent findings in molecular biology shed light on the processes of evolution. Wills begins by discussing some of the problems and challenges that face evolutionary biologists. He believes that evolution is getting easier with time due to "the way that genes have come to be arranged in organisms and by the evolving nature of the factors that alter them." After condemning the creation science movement's attempt to force the teaching of creationism in public schools, he cautions readers against some evolutionary fallacies: that organisms are directing their own evolution, that evolution is goal directed, that species usually evolve smoothly into new species, and that the fossil record provides a complete picture of evolution. He supports the concepts of punctuated equilibrium and species selection and examines how mutation, jumping genes, homeotic genes, domain shuffling in proteins, and other genetic phenomena facilitate evolutionary change. Wills believes that a growing understanding of how evolution works is beginning to allow scientists to influence the direction of the evolutionary process through gene manipulation.

Author Index

This Author Index contains only the names of authors whose works are annotated in this bibliography. The names of other authors mentioned in the annotations are included in the Subject Index.

AUTHOR INDEX

TITLE INDEX

This Title Index contains only the titles of works annotated in this bibliography. The names of other titles mentioned in the annotations are included in the Subject Index.

TITLE INDEX

Subject Index

This Subject Index does not include the names of books or the authors of books annotated in this bibliography, except for those mentioned within other annotations. Included book names are italicized and followed by the name(s) of the author(s) in parentheses.

About the Author

James L. Hayward earned a Ph.D. in zoology at Washington State University in 1982. He is a professor of biology at Andrews University, Berrien Springs, Michigan, where he teaches courses in genetics, ecology, and the history of life. His research in Washington, Montana, and Alberta involves characterizing the nesting ecology of ancient dinosaurs and living gulls.